C0-AZG-710

SATELLITE REMOTE SENSING of NATURAL RESOURCES

MAPPING SCIENCES SERIES

John G. Lyon, Series Editor

Aerial Mapping: Methods and Applications
Edgar Falkner, U.S. Army Corps of Engineers

Environmental GIS: Applications to Industrial Facilities
William J. Douglas, Environmental Resources Management, Inc.

Practical Handbook for Wetland Identification and Delineation
John G. Lyon, Ohio State University

Satellite Remote Sensing of Natural Resources
David L. Verbyla, University of Alaska, Fairbanks

Wetland and Environmental Engineering Applications of GIS
John Lyon, Ohio State University

SATELLITE REMOTE SENSING of NATURAL RESOURCES

David L. Verbyla

Library of Congress Cataloging-in-Publication Data

Verbyla, David
Satellite remote sensing of natural resources / David Verbyla.
p. cm.
Includes bibliographical references and index.
ISBN 1-56670-107-4
1. Remote sensing. I. Title.
G70.4.V47 1995
621.36'78—dc20 94-23519
CIP

Direct all inquiries to CRC Press, Inc., 2000 Corporate Blvd., N.W., Boca Raton, Florida 33431.

Lewis Publishers is an imprint of CRC Press

International Standard Book Number 1-56670-107-4
Library of Congress Card Number 94-23519
Printed in the United States of America 1 2 3 4 5 6 7 8 9 0
Printed on acid-free paper

Mapping Sciences Series

The Mapping Sciences Series of books was conceived to serve important needs in the profession. Readers require contemporary information of a practical value, with high-quality theoretical detail. They need to know methods and approaches for their given application. Authors have longed for a coherent series that addresses the variety of disciplines and applications that come under the umbrella of mapping sciences. They require an outlet for their expertise and a series that can be readily identified by their audience. The combination of these needs and their solution has yielded the Mapping Sciences Series.

John G. Lyon
Editor-in-Chief
Lewis Mapping Sciences Series
and Associate Professor
of Civil Engineering
Ohio State University

Preface

This is a simple book. It is truly introductory, and assumes no statistical or mathematical background beyond an elementary statistics course. The goal of this primer is to help students understand the basics of digital remote sensing, and to bridge the gap to more advanced texts. Since this is an introductory primer, basic concepts are emphasized. You will learn many of these concepts by simply using a ruler to plot relationships on paper. These concepts can be extended to n-dimensions. However, because the goal is simplicity, I usually restrict presentation to two-dimensions. To learn more about the topics presented, I encourage you to read the Additional Readings citations at the end of each chapter. After you read each chapter, test your understanding by trying the problems at the end of the chapter. The solutions to all even-numbered problems are presented in Appendix A.

This primer has evolved from my teaching of courses and workshops to natural resource students and managers at four different universities in the United States. These students and managers are typically in renewable natural resources: forestry, wildlife management, range science, soil science, watershed management, environmental planning, and related disciplines. These students often have a limited math/stat background and are interested in practical applications of satellite remote sensing, especially as GIS applications in natural resources. If you are like these students, I hope this primer will help you.

I thank the many students and managers who offered constructive criticism of drafts of this primer including Barbara Boyle, Song Choung, Carl Markon, Carla Richardson, and Ken Winterberger. I especially thank Joe Ulliman and Ahmed Fahsi for reviewing the entire first draft of the manuscript, and Tim Hammond and Scott Rupp for reviewing the entire second draft.

Dave Verbyla
Fairbanks, Alaska

The Author

David L. Verbyla, Ph.D., is an Assistant Professor of GIS/Remote Sensing in the Department of Forest Sciences, School of Agriculture and Land Resources Management, at the University of Alaska, Fairbanks. He has taught remote sensing/GIS courses or workshops at Utah State University, the University of New Hampshire, the University of Idaho, and the University of Alaska, Fairbanks.

Dr. Verbyla's research interests focus on practical applications of remote sensing and GIS for natural resources management, including validation of models, classification accuracy assessment, wildfire mapping, classification of boreal forests, monitoring of sediment sources, and mapping of climatic zones. Dr. Verbyla spends his spare time enjoying Alaska's wonderful natural resources.

Contents

CHAPTER 1

Satellite Images

RASTER IMAGE DATA

A digital satellite image consists of a grid of numbers (Figure 1.1). Each grid cell is called a *pixel* (or picture element). The values inside each pixel are called *digital numbers* (DNs) or data file values. These grid values may represent quantities (such as elevation, slope gradient, or spectral brightness values) and are often called digital images. Cell values sometimes represent predicted classes or categories such as vegetation type (e.g., 1 = aspen, 2 = balsam poplar, 3 = white spruce) or forest crown closure class (e. g., 1 = less than 30%, 2 = 30 to 60%, 3 = greater than 60%) and are sometimes called classified images. One goal of satellite remote sensing is to process an original digital image into an accurate and useful classified image.

Data organized in a grid cell format are called raster data (from the German word for rake, since scanning systems utilize a raking motion). Most raster systems store information in a column, row or X,Y format with the origin (column 1, row 1) occurring at the upper left hand corner of the image file.

REMOTE SENSING DETECTORS

Remote sensing detectors are electronic devices that are sensitive to radiation. Several different detectors are needed to cover a wide range of wavelengths. For example, Landsat Thematic Mapper uses silicon detectors (sensitive in the visible and near-infrared spectral region), indium-antimonide detectors (sensitive in the mid-infrared spectral region), and mercury-cadmium-telluride detectors (sensitive in the thermal infrared spectral region). SPOT HRV uses a row of thousands of silicon detectors and therefore SPOT HRV images are are currently restricted to spectral bands from the visible and near-infrared spectral regions. Detectors respond by producing a voltage signal (the higher the radiation detected, the higher the detector voltage response). For example, imagine that we have four silicon detectors that are sensitive to light wavelengths ranging from

	COL #1	COL #2	COL #3	COL #4	COL #5
ROW #1:	15	23	29	32	35
ROW #2:	17	24	25	25	32
ROW #3:	19	27	23	23	30

Figure 1.1. Example of a simple 5-column by 3-row raster digital image.

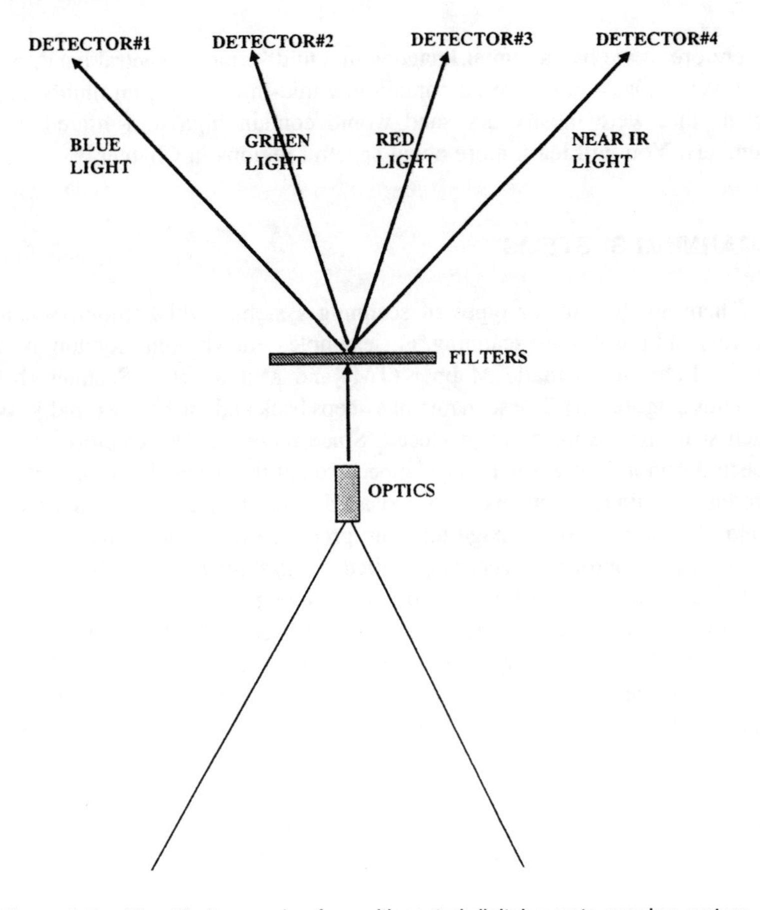

Figure 1.2. Simplified example of a multispectral digital remote sensing system.

0.3 to 1.0 μm. We can place interference filters over the detectors so that each detector "sees" only blue, green, red, or near-infrared radiation (Figure 1.2).

Analog to Digital Conversion

As the sensor system travels over different land cover types, voltage response from each sensor will vary. For example, as the sensor passes over highly reflective sand, detector response voltage will be high compared to when the sensor passes over highly absorbtive water. These analog voltage responses are converted to digital numbers by sampling the sensor voltages at a fixed interval. This process is extremely fast; Landsat Thematic Mapper samples at a rate of 9.6 millionth of a second! High digital numbers will correspond to ground pixel areas that had high reflectance for that spectral band. For example, water reflects very little mid-infrared radiation, while dry soil reflects more of the incident radiation in the mid-infrared spectral region.

Therefore, if we had a digital image with a mid-infrared spectral band, pixels that were mostly water would contain low mid-infrared digital numbers, and pixels that were mostly dry sand would contain high mid-infared digital numbers. You will learn more about spectral regions in Chapter 3.

SCANNING SYSTEMS

There are two major types of scanning systems: whiskbroom (scanning mirror) and pushbroom scanning. For example, whiskbroom scanning is used in the Landsat Thematic Mapper (TM) and Multispectral Scanner (MSS) systems (Figure 1.3). The scan mirror sweeps back and forth very rapidly. With each scan, image lines are produced. Since there are 16 detectors for each spectral region in the Thematic Mapper (except the thermal band), each scan produces 16 lines of image data. Landsat MSS has 6 detectors in each spectral band; therefore, 6 MSS image lines are produced from each scan.

A second approach to scanning, called pushbroom scanning, is used in the SPOT satellites (Figure 1.4). Pushbroom scanning is a newer technology than whiskbroom scanning. A row of silicon detectors, called a charge coupled device (CCD) linear array, is used. SPOT HRV uses a row of 6000 detectors when operating in the panchromatic (a single wide spectral band with 10 meter-square ground pixels) mode, and 3000 detectors in the multispectral (3 spectral bands with 20 meter-square ground pixels) mode. The advantages of pushbroom scanning include: 1) elimination of geometric errors due to variations in the mirror scan velocity, 2) fewer moving parts (and therefore a longer life expectancy), and 3) increased dwell time to sample radiation within each ground

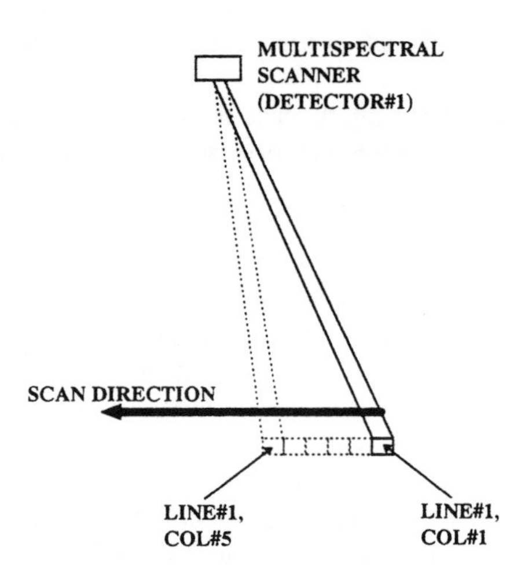

Figure 1.3. Whiskbroom scanning used by Landsat Multispectral Scanner and Thematic Mapper scanning systems.

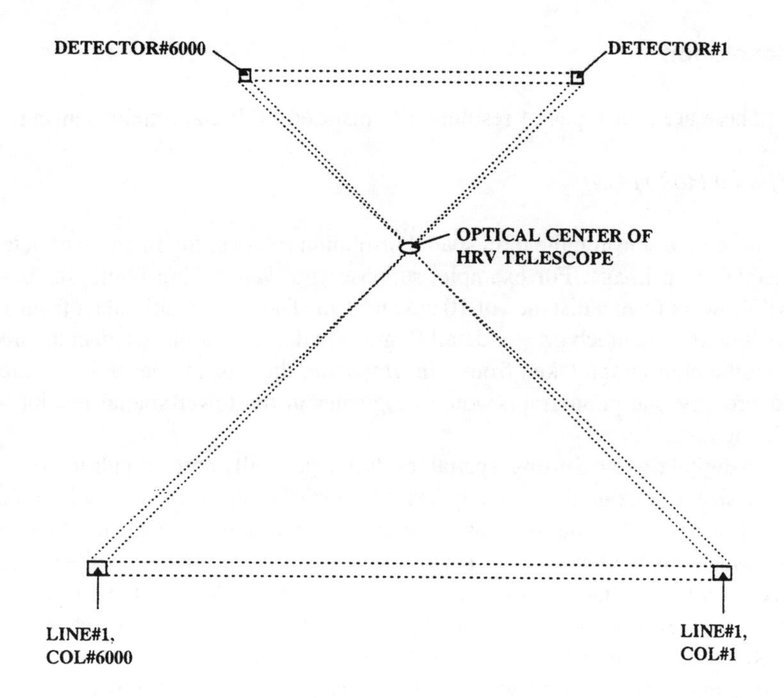

Figure 1.4. Pushbroom scanning used by SPOT scanning system.

pixel. A disadvantage of pushbroom scanning in SPOT satellites is that because silicon detectors are used, the spectral range of pushbroom scanning currently does not include mid-infrared or thermal infrared.

IMAGE SCALE AND RESOLUTION

Scale

Sometimes students confuse spatial resolution with image scale. Scale can be defined as the number of ground units represented by a single unit on an image. For example, we could print a SPOT panchromatic image showing clearcut areas at a scale of 1:50,000. This would mean that an inch on the image represents 50,000 inches on the ground, a cm on the image represents 50,000 cm on the ground, a mm on the image represents 50,000 mm on the ground, and so on.

We could print the same image on a smaller piece of paper (and thus have a smaller scaled image). For example, we could print the image at a scale of 1:75,000 or 1:100,000. Scale is *independent* of pixel size! For example, we could print an image out at a scale of 1:25,000 or 1:50,000 or 1:75,000 or 1:100,000 and use an image with 10-meter pixels for each of the printed images.

Resolution

There are four types of resolution terms used in digital remote sensing.

Spatial Resolution

In remote sensing, the term spatial resolution refers to the fineness of detail visible in an image. For example, suppose you were taking photographs of wildflowers from a distance of 10 cm and 1 m. The photograph taken from 10 cm would have much greater detail (higher spatial resolution) when compared with the photograph taken from 1 m. However, there is a tradeoff — the area covered by one photograph would be greater in the lower spatial resolution photograph.

In digital remote sensing, spatial resolution generally corresponds to ground pixel size. For example, you may hear that SPOT has a 10-meter resolution or Landsat Thematic Mapper has a 30-meter resolution. This means that the instantaneous field of view the sensor "sees" (which corresponds to the ground pixel size) is a square with 10-meter or 30-meter sides. You will learn that this does not necessarily mean that objects less than the pixel size cannot be detected. For example, it is possible to detect 10-meter wide roads on Landsat Thematic Mapper images when there is a sharp contrast between roads and the surrounding areas such as a dry, logging road surrounded by dark, dense forest.

There is a tradeoff with increased spatial resolution. As the spatial resolution increases, so does the necessity for more computer disk space; this results in greater disk storage and processing costs. For example, there are nine times more pixels covering a given area in a 10-meter pixel image than there are in a 30-meter pixel image.

Spectral Resolution

Spectral resolution refers to the width across the electromagnetic spectrum that the remote sensing instrument is detecting. For example, SPOT panchromatic images are recorded with lower spectral resolution (a wide spectral band) when compared with Landsat Thematic Mapper images (many relatively narrow spectral bands ranging from blue to thermal infrared). Landsat Thematic Mapper, with better spectral resolution, is superior to SPOT panchromatic data for vegetation mapping because many vegetation types can be delineated due to spectral differences. This is analogous to trying to visually interpret coniferous forest versus broadleaf forest using superior filtered color infrared photography versus unfiltered panchromatic photography.

Radiometric Resolution

Radiometric resolution refers to the ability of a remote sensing system to record many levels of values. For example, Landsat MSS data were recorded in grid cell values ranging from 0 to 63 and therefore have lower radiometric

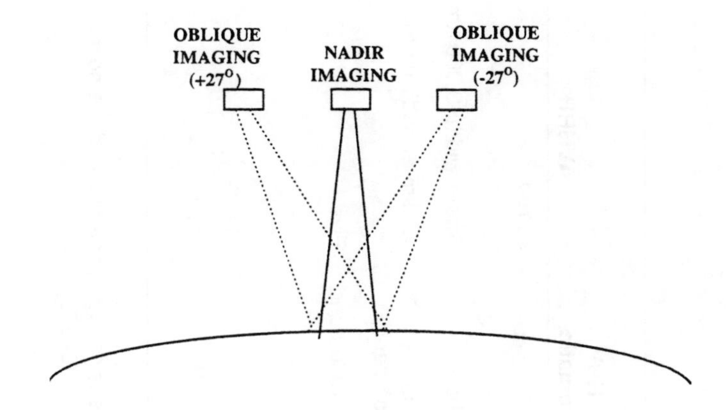

Figure 1.5. Off-nadir viewing capability of SPOT using pointable mirror.

resolution when compared with Landsat Thematic Mapper data which are recorded in a range from 0 to 255. An analogy would be a grading system of A, B, C, F versus a higher resolution system of A+, A, A–, B+, B, B–, C+, C, C–, D, F.

Temporal Resolution

Temporal resolution is the imaging revisit interval. In terms of vertical images (images taken at nadir), Landsat has higher temporal resolution compared to SPOT since the orbit cycle for Landsat is 16 days while the orbit cycle for SPOT is 26 days. However, SPOT has a pointable mirror which allows for oblique viewing (Figure 1.5). Therefore, because of the capability of oblique or off-nadir imaging, SPOT has higher temporal resolution when compared with Landsat (which is capable only of nadir viewing).

Temporal resolution also generally increases at higher latitudes due to significant sidelap between consecutive satellite passes. For example, at the equator the same area can be imaged 8 or 9 times during a SPOT 26-day orbit cycle. At 45° latitude, an area can be imaged 12 times per 26-day orbit cycle. The percent sidelap of consecutive Landsat scenes also varies with latitude, ranging from 15% at the equator to 85% at extreme latitudes.

MAJOR SATELLITE SYSTEMS USED IN NATURAL RESOURCES MANAGEMENT

Each satellite system has advantages and disadvantages; the most appropriate satellite imagery depends on the objectives of the natural resource manager. Some characteristics of satellite imagery are shown in Table 1.1.

Table 1.1. Comparison of Landsat, SPOT, and AVHRR digital scenes.

	LANDSAT MSS	LANDSAT Thermatic Mapper	SPOT HRV Multispectral	SPOT HRV Panchromatic	AVHRR
1994 Cost per scene	$1,000[a]	$4,900	$2,450	$2,450	<$100
Scene area	3.3 million ha	3.1 million ha	360,000 ha	360,000 ha	Swath width: 3,000 km
Pixel size	56 by 79 m	30 by 30 m	20 by 20 m	10 by 10 m >6 km off-nadir	1.1 km at nadir up to
Number of spectral bands	Four	Seven	Three	One very wide band	Five
Radiometric resolution	127 levels (7-bit)	255 levels (8-bit)	255 levels (8-bit)	255 levels[b] (8-bit)	1024 (10-bit)
Year of first images	1972	1982	1986	1986	1979

[a] Landsat MSS data older than 2 years are available at $200 per scene

[b] SPOT panchromatic data are originally coded as DPCM code where the first value is coded over 8-bits and the next two are coded over 5-bits. This information is latter expanded so that all pixels are 8-bit (0–255).

Landsat

Landsat utilizes two sensor systems: Landsat Multispectral Scanner (MSS) and Landsat Thematic Mapper (TM). MSS data are available since 1972 on an orbit cycle of every 16 to 18 days. TM data are available since 1982 on an orbit cycle of every 16 days. There have been five Landsat satellites in operation, starting with Landsat-1 in 1972. Currently Landsat-5 is operational and records both MSS and TM data with an equatorial crossing time of 11 AM. Landsat MSS data are of lower spatial and spectral resolution than TM data. However, MSS data have two advantages over TM data: 1) their longer history (since 1972) makes them useful for change detection studies such as forest fragmentation, fire history, etc.), and 2) they are much less expensive and require less computer storage space than TM data.

SPOT

The SPOT (Le Systeme Pour l' Observation de la Terre) satellite has two identical high-resolution visible (HRV) sensors. The HRV sensors can operate in two modes: a *panchromatic* mode and a *multispectral* mode. A panchromatic photograph is a "black and white" photograph usually produced by film chemicals that are sensitive to a wide range of visible wavelengths. SPOT panchromatic images are analogous to panchromatic photographs; the images are the result of sensors that received reflected radiation composed of a wide range of visible wavelengths. Panchromatic data, with 10-meter ground pixels, have high spatial resolution and low spectral resolution. Panchromatic imagery would be a poor choice for vegetation mapping because of the poor spectral resolution. Consider this analogy: you would prefer color or color-infrared photography instead of panchromatic photography for delineating spruce versus aspen forests, because the aspen and spruce forests would appear as similar shades of grey on the panchromatic photographs due to lower spectral resolution. Because of improved spectral resolution, the aspen and spruce stands would appear as distinctly different colors on either the color or color infrared photographs. However, because of relatively high spatial resolution, SPOT panchromatic imagery would be an excellent choice for mapping of roads, clearcut boundaries, rock outcrops, water bodies, and other high-contrast features. Think of this analogy: because of higher spatial resolution, you would prefer large-scale panchromatic aerial photographs instead of small-scale color photographs for mapping hiking trails and other small, high-contrast features.

Multispectral SPOT data are from three bands within the green, red, and near-infrared spectral regions. These spectral bands are narrower than the SPOT panchromatic band (Table 1.1), and therefore, have higher spectral resolution compared to SPOT panchromatic data. SPOT multispectral images have relatively high spatial resolution (20-meter pixels). However, SPOT multispectral data currently lack mid-infrared bands which may be important for mapping moisture or shadow-related features. SPOT data also lack a thermal band, which is critical in some applications such as mapping lake water temperature zones.

The SPOT satellite system is unique in that it has pointable mirrors which allow for flexible user-specified ground coverage. This high temporal resolution is important for monitoring dynamic events such as forest fires, flooding, vegetation phenology, etc. If the mirrors were not pointable, the revisit frequency of SPOT would be only every 26 days. Because the pointable mirrors allow images to be acquired at different angles, overlapping SPOT satellite images can be viewed with a stereoscope in three dimensions (similar to aerial photographs). Because of this stereo-viewing capability, SPOT images can be used to produce digital elevation models for remote terrain areas worldwide.

AVHRR

The Advanced Very High Resolution Radiometer (AVHRR) sensors are on National Oceanic and Atmospheric Administration (NOAA) weather satellites. AVHRR was originally designed for imaging cloud cover and has been adopted for coarse scale (1-km pixel at best) terrestrial applications. AVHRR images have high panoramic distortion with a pixel size of approximatly 1-km at nadir and over 6-km towards the beginning and end of each scan line. Because of the low spatial resolution, AVHRR data are sometimes not very useful for local natural resource applications. However, for large area, high temporal resolution applications, AVHRR data are very useful. Because AVHRR data are acquired on a 12-hr cycle, cloud cover is less of a problem compared to lower temporal resolution satellites. Also, because AVHRR data are commonly stored in 1-km pixel size, the volume of data from large areas is relatively small compared to higher spatial resolution data. For example, there are more than 1,100 Landsat Thematic Mapper 30-meter pixels per 1 AVHRR 1-km pixel. Therefore, the analysis of AVHRR data for a given area would be much quicker compared to the analysis of higher spatial resolution data. It is possible to remotely detect subpixel phenomena such as wildfires using AVHRR data because there can be a great contrast in AVHRR channel 3 between pixels containing fires and pixels not containing fires.

CHAPTER 1 PROBLEMS

1) Calculate the cost per hectare of the following:

   1 Landsat MSS scene:
   1 Landsat Thematic Mapper Scene:
   1 SPOT Panchromatic Scene:

   Landsat Thematic Mapper data are available for limited dates on floppy disks for $600. The seven-band image consists of 512 rows by 512 columns. Calculate the cost per hectare of this image:

SPOT data can be purchased as "GIS Digital Quadmaps" corresponding in area to the U.S. Geological Survey 7.5 minute quadrangle maps. The price for a digital quad map is $950. Select a 7.5-minute quadrangle map of your home town. Compute the cost per hectare of the digital quadmap for the same area.

2) Suppose you have a Landsat Thematic Mapper image consisting of 1000 rows by 1000 columns. If you printed this image on a piece of paper that was 100 cm by 100 cm in size, what would the scale of the printed image be?

3) Select whether aerial photography or satellite imagery would be more appropriate for the following applications (discuss your reasoning) :

   A) Mapping of Sitka spruce stands in coastal Pacific Northwest area where cloud cover is a problem 90% of the time.
   B) Mapping new hiking trails that have recently been established in a park.
   C) Analyzing elk habitat based on forest stand height class and crown density class.
   D) Analyzing 5-year interval loss of wetlands in a county since 1960.
   E) Mapping of waterfowl habitat within the intertidal zone of an estuary.

4) Match the following resolution terms with the best analogy (list one letter for each resolution term):

_____ Higher radiometric resolution
_____ Lower radiometric resolution
_____ Higher spatial resolution
_____ Lower spatial resolution
_____ Higher spectral resolution
_____ Lower spectral resolution
_____ Higher temporal resolution
_____ Lower temporal resolution

   A) Taking a color photograph without using a filter.
   B) Taking a color photograph with a yellow (minus blue) filter.
   C) Subscribing to a daily newspaper.
   D) Subscribing to a monthly magazine.
   E) Using a pop-up indicator to check whether a roasted turkey is done.
   F) Using a meat thermometer to check whether a roasted turkey is done.
   G) Taking a photograph of an elk herd with a 25-mm wide-angle lens.
   H) Taking a photograph of an elk with a 400-mm zoom lens.

ADDITIONAL READINGS

General Readings

Duggin, M. J. and C. J. Robinove. 1990. Assumptions implicit in remote sensing data acquisition and analysis. *International Journal of Remote Sensing.* 11:1669–1694.

Glick, D. 1994. Windows on the world: satellite and shuttle remote sensing of earth. *National Wildlife.* 32:4–13.

Jones, R. C. 1968. How images are detected. *Scientific American.* 219:111–117.

Landgrebe, D. 1983. Land observation sensors in perspective. *Remote Sensing of Environment.* 13:391–402.

Light, D. 1990. Characteristics of remote sensors for mapping and earth science applications. *Photogrammetric Engineering and Remote Sensing.* 56:1613–1625.

Philipson, W. R. 1986. Problem-solving with remote sensing: an update. *Photogrammetric Engineering and Remote Sensing.* 52:109–110.

Sayn-Wittgenstein, L. 1992. Barriers to the use of remote sensing in providing environmental infomation. *Environmental Monitoring and Assessment.* 20:159–166.

Satellite Remote Sensing of Natural Resources

Botkin, D. B., Estes, J. E., MacDonald, R. M., and M. V. Wilson. 1984. Studying the earth's vegetation from space. *BioScience.* 34:508–514.

Colwell, R. N. 1967. Remote sensing as a means of determining ecological conditions. *BioScience.* 17:444–449.

Dottavio, C. L. and D. L. Williams. 1983. Satellite technology: an improved means for monitoring forest insect defoliation. *Journal of Forestry.* 81:30–34.

Gates, D. M. 1967. Remote sensing for the biologist. *BioScience.* 17:303–307.

Greegor, D. H. 1986. Ecology from space. *BioScience.* 36:429–432.

Hall, F. G., Botkin, D. B., Strebel, D. E., Woods, K. D. and S. J. Goetz. 1991. Large-scale patterns of forest succession as determined by remote sensing. *Ecology.* 72:628–640.

Iverson, L. R., Graham, R. L. and E. A. Cook. 1989. Applications of satellite remote sensing to forested ecosystems. *Landscape Ecology.* 3:131–143.

Leckie, D. G. 1990. Advances in remote sensing technologies for forest surveys and management. *Canadian Journal of Forest Research.* 20:464–483.

Meyer, M. P. and L. F. Werth. 1990. Satellite data: management panacea or potential problem? *Journal of Forestry.* 88(9):10–13.

Moore, M. M. 1990. Remote sensing and geographical information systems for integrated resource management: an overview. In: *Proceedings of Integrated Management of Watersheds for Multiple Use.* pp. 79–89. USDA Forest Service General Technical Report. RM-198.

O'Neill, T. 1993. New sensors eye the rain forest. National Geographic. 184:118.

Roughgarden, J., Running, S. W. and P. A. Matson. 1991. What does remote sensing do for ecology? *Ecology.* 72: 1918–1922.

Trotter, C. M. 1991. Remotely-sensed data as an information source for geographical information systems in natural resource management: a review. *International Journal of Geographical Information Systems.* 5:225–239.

Tueller, P. T. 1992. Overview of remote sensing for range management. *Geocarto International.* 7:5–10.

Landsat Readings

Homes, R. A. 1984. Advanced sensor systems: thematic mapper and beyond. *Remote Sensing of Environment.* 15:213–221.

Markham, B. L. and J. L. Barker. 1985. Spectral characterization of the Landsat Thematic Mapper sensors. *International Journal of Remote Sensing.* 6:697–716.

Phillips, K. M., Morgan, K., Newland, L. and D. G. Koger. 1986. Thematic mapper data: A new land planning tool. *Journal of Soil and Water Conservation.* 41:301–303.

Schwaller, M. and B. Dealy. 1986. Landsat and GIS. *Journal of Forestry.* 84:40–41

U.S. Geological Survey. 1984. *Landsat 4 Data Users Handbook.* USGS, Alexandria, VA. 80 pp.

SPOT Readings

Arnaud, M. and M. Leroy. 1991. SPOT 4: a new generation of SPOT satellites. *ISPRS Journal of Photogrammetry and Remote Sensing.* 46:205–215.

Chevrel, M., Courtois, M., and G. Weill. 1981. The SPOT satellite remote sensing mission. *Photogrammetric Engineering and Remote Sensing.* 47:1163–1171.

SPOT Image Corp. 1988. *SPOT User's Handbook.* SPOT Image Corp., Reston, VA. 2 vols.

Thompson, L. L. 1979. Remote sensing using solid-state array technology. *Photogrammetric Engineering and Remote Sensing.* 45:47–55.

AVHRR Readings

Achard, F. and F. Blasco. 1990. Analysis of vegetation seasonal evolution and mapping of forest cover in West Africa with the use of NOAA AVHRR HRPT data. *Photogrammetric Engineering and Remote Sensing.* 10:1359–1365.

Barber, D. G. and P. R. Richard. 1992. Use of AVHRR imagery in Arctic marine mammal research. *International Journal of Remote Sensing.* 13:167–175.

Ehrlich, D., Estes, J. E. and A. Singh. 1994. Applications of NOAA-AVHRR 1 km data for environmental monitoring. *International Journal of Remote Sensing.* 15:145–161.

Eidenshink, J. C. 1992. The conterminous U. S. AVHRR data set. *Photogrammetric Engineering and Remote Sensing.* 809–813.

Ferris, D. S. and R. G. Congalton. 1989. Satellite and geographic information system estimates of Colorado River basin snowpack. *Photogrammetric Engineering and Remote Sensing.* 55:1629–1635.

Iverson, L. R., Cook, E. A. and R. L. Graham. 1989. A technique for extrapolating and validating forest cover across large regions. Calibrating AVHRR data with TM data. *International Journal of Remote Sensing.* 10:1805–1812.

Kidwell, K. B. (ed.) 1991. *NOAA Polar Orbiter Data Users Guide.* NOAA, National Climatic Data Center, Satellite Data Service Division, Washington, D. C. 184 pp.

Lopez, S., Gonzales, F., Llop, R, and J. M. Cuevas. 1991. An evaluation of the utility of NOAA AVHRR images for monitoring forest fire risk in Spain. *International Journal of Remote Sensing.* 12: 1841–1851.

Malingreau, J. P., Tucker, C. J. and N. Laporte. 1989. AVHRR for monitoring global tropical deforestation. *International Journal of Remote Sensing.* 10:855–867.

Pereira, M. C. and A. W. Setzer. 1993. Spectral characteristics of deforestation fires in NOAA/AVHHR images. *International Journal of Remote Sensing.* 14:583–597.

Robinson, J. M. Fire from space: global fire evaluation using infrared remote sensing. *International Journal of Remote Sensing.* 12:3–24.

Spanner, M. A., Pierce, L. L., Running, S. W., and D. L. Peterson. 1990. The seasonality of AVHRR data of temperate coniferous forests: relationships with leaf area index. *Remote Sensing of Environment.* 33:91–112

Tappan, G. C., Moore, D. G., and W. I. Knausenberger. 1991. Monitoring grasshopper and locust habitats in Sahelian Africa using GIS and remote sensing technology. *International Journal of Remote Sensing.* 5:123–135.

Tueber, K. B. 1990. Use of AVHRR imagery for large-scale forest inventories. *Forest Ecology and Management.* 33/34:621–631.

CHAPTER 2

Image Processing Systems

INTRODUCTION

Since satellite images are digital, they can be processed efficiently by computer systems called image processing systems. Image processing systems consist of specialized hardware and software for the analysis and display of digital images. Such systems vary from very expensive and powerful mainframe systems to relatively inexpensive microcomputer systems. These systems often have "user-friendly" menus. However, to make the right selections from the menus, you should understand the fundamentals of image processing systems.

COMPUTER FUNDAMENTALS

Most computers consist of three basic components: 1) input/output units, 2) data processing units, and 3) data storage units. Common input units include the keyboard, mouse, joystick, and digitizer tablet. Output units commonly used in digital remote sensing include a color monitor and color printer or plotter.

The data processing unit is the "brain" of a computer called the central processing unit (often referred to as the CPU). The CPU usually has 2 main jobs: 1) to perform arithmetic and logic operations, and 2) to control the flow of data into and out of other computer components. Some computer systems also have an efficient math coprocessor to relieve the main CPU from the burden of performing real number (floating point) calculations. In fact, many microcomputer-based image processing systems and geographic information systems require a math coprocessor.

Computers usually have two types of data storage units: internal data storage (memory) and external storage (such as disks and tape drives). The advantage of memory is that it uses integrated electric circuits for processing and is therefore extremely fast. However, memory is expensive for data storage and it is usually temporary (data are lost from memory when the electricity is shut off). External storage, such as hard disks and tape drives, can store massive amounts of data permanently and relatively inexpensively. However, data access on external storage devices is slow relative to internal memory data access speeds.

Bits and Bytes

All computer data — integers, real numbers, letters, special characters, etc. — are stored in bit format. The term *bit* stands for binary digit, and can represent only two states: on and off. For example, we might have one bit (or memory cell) that represents the number zero when it is off, and the number 1 when it is on:

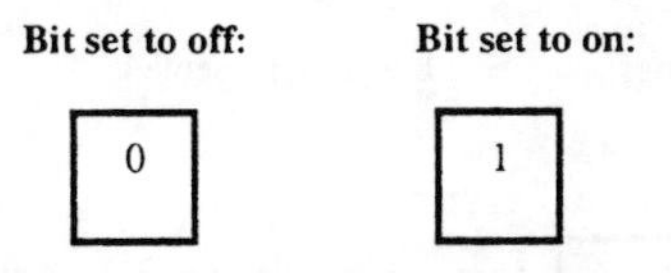

Figure 2.1. Binary digit on/off settings.

Now imagine that we have 2 bits instead of 1. How many integers could we represent?

Bit settings: 2^1:	2^0:	Integer represented:
0	0	0 $(0 \times 2^1 + 0 \times 2^0 = 0 + 0)$
0	1	1 $(0 \times 2^1 + 1 \times 2^0 = 0 + 1)$
1	0	2 $(1 \times 2^1 + 0 \times 2^0 = 2 + 0)$
1	1	3 $(1 \times 2^1 + 1 \times 2^0 = 2 + 1)$

Figure 2.2. Integers represented by 2-bit data storage.

Since we have 2 bits, we could represent 22 or 4 different integers (0,1,2, or 3). Using the same reasoning, if we have 3 bits, we could represent 23 or $2 \times 2 \times 2 = 8$ different integers. These 8 integers could be represented as shown in Figure 2.3.

Satellite data often come in 8-bit format. How many different integers can be represented with 8 bits (Figure 2.4)?

A bit of trivia (no pun intended): 4 bits are called a nibble, and 8 bits are called a byte. Computer memory and disk storage capacity is often expressed in megabytes (1 million bytes) and gigabytes (1 billion bytes). One byte can represent any integer from 0 to 255. Alphanumeric and special characters can be represented by one byte using a special coding scheme called ASCII format (Table 2.1). ASCII stands for the American Standard Coding Information Interchange and is a standard format used to transfer data among computers.

Magnetic Tapes

Satellite data are often purchased on 9-track and 8-mm magnetic tapes. A 9-track tape has 9 tracks; each track can hold a stream of 1s and 0s. Since a byte is composed of 8 bits, the first 8 tracks store sequential bytes of data. The ninth track holds a special bit, called a parity bit, and is used for error checking

Bit settings: 2^2:	2^1:	2^0:	Integer represented:
0	0	0	0 $(0 X 2^2 + 0 X 2^1 + 0 X 2^0 = 0 + 0 + 0)$
0	0	1	1 $(0 X 2^2 + 0 X 2^1 + 1 X 2^0 = 0 + 0 + 1)$
0	1	0	2 $(0 X 2^2 + 1 X 2^1 + 0 X 2^0 = 0 + 2 + 0)$
0	1	1	3 $(0 X 2^2 + 1 X 2^1 + 1 X 2^0 = 0 + 2 + 1)$
1	0	0	4 $(1 X 2^2 + 0 X 2^1 + 0 X 2^0 = 4 + 0 + 0)$
1	0	1	5 $(1 X 2^2 + 0 X 2^1 + 1 X 2^0 = 4 + 0 + 1)$
1	1	0	6 $(1 X 2^2 + 1 X 2^1 + 0 X 2^0 = 4 + 2 + 0)$
1	1	1	7 $(1 X 2^2 + 1 X 2^1 + 1 X 2^0 = 4 + 2 + 1)$

Figure 2.3. Integers represented by 3-bit data storage.

during data transmission. The density of data stored on a 9-track tape is often described in terms of bits per inch. For example, a 1600 bpi tape would store 1600 bits per inch, while a 6250 bpi tape would store 6,250 bits per inch of tape. Because 8-mm tapes store data at such a high density, one tape can hold more than a gigabyte (1 billion bytes) of data! Magnetic tapes are often used to store backup copies of data because of their massive data storage capacity and relatively inexpensive cost. Hundreds of tapes can be maintained in a library to provide a well-organized backup system of valuable data and software.

Although tapes can store massive quantities of data, tape storage systems have the disadvantage of retrieving data at a relatively slow speed. This is because the data are stored sequentially on the tape. Therefore, to access a file in the middle of a tape, the tape drive has to physically skip over all files from

2^7:	2^6:	2^5:	2^4:	2^3:	2^2:	2^1:	2^0:	
0	0	0	0	0	0	0	0	= 0
1	1	1	1	1	1	1	1	= 255

Figure 2.4. Range of integers represented by 8-bit data storage.

Table 2.1. ASCII character set

Decimal Value	ASCII Character	Decimal Value	ASCII Character	Decimal Value	ASCII Character
0–31	Special chars	64	@	97	a
32	Space	65	A	98	b
33	!	66	B	99	c
34	“	67	C	100	d
35	#	68	D	101	e
36	$	69	E	102	f
37	%	70	F	103	g
38	&	71	G	104	h
39	‘	72	H	105	i
40	(	73	I	106	j
41	)	74	J	107	k
42	*	75	K	108	l
43	+	76	L	109	m
44	‘	77	M	110	n
45	-	78	N	111	o
46	.	79	O	112	p
47	/	80	P	113	q
48	0	81	Q	114	r
49	1	82	R	115	s
50	2	83	S	116	t
51	3	84	T	117	u
52	4	85	U	118	v
53	5	86	V	119	w
54	6	87	W	120	x
55	7	88	X	121	y
56	8	89	Y	122	z
57	9	90	Z	123	{
58	:	91	[	124	|
59	;	92	\	125	}
60	<	93	]	126	~
61	=	94	^	127	Delete
62	>	95	_	128–255	Special chars
63	?	96	‘		

the beginning to the middle of the tape. This is analogous to trying to listen to a song located at the middle of a cassette tape — you must first push the fast forward button to skip over all songs from the beginning of the cassette tape until you reach the song at the middle of the tape.

Packaging Multispectral Data

Multispectral image data are usually packaged on tapes and disks in one of three formats: 1) band sequential (BSQ), 2) band interleaved by line (BIL), or

3) band interleaved by pixel (BIP). Imagine that we have the following 6-pixel, 3-band image:

56	58	62	BAND 1
69	82	94	BAND 2
134	135	129	BAND 3
148	197	152	BAND 1
156	157	143	BAND 2
120	172	184	BAND 3

Figure 2.5. Six-pixel, 3-band digital image.

The data from Figure 2.5 could be packaged either as BSQ, BIL or BIP (Figure 2.6). Therefore it is important to know how the image data are packaged. A common mistake is reading an image in using the wrong packaging format — the image then appears as random noise when displayed to the monitor screen.

Tape header files typically contain information such as number of rows and columns in the image, number of spectral bands, sensor information, and other descriptive information about the data. Some image-processing systems store tape header information as records at the start of the image file. Other systems store header information as a companion file with each image file.

DISPLAY OF PANCHROMATIC IMAGES

Digital images can be displayed on a computer monitor by using the pixel values to control the video intensity of each pixel on the screen. Usually the intensity will range from 0 (very dark) to 255 (very bright). For example, we might have an image of a river bordered by sandy shorelines. Such an image could be displayed as shown in Figure 2.7.

Imagine we had an image of a marsh area where the area is composed of open water and bulrush marsh. The digital image might be as shown in Figure 2.8.

How would this image look when it is displayed on a computer monitor?

The problem is that we have potential video intensities ranging from 0 to 255; however, by using the data to control the video intensities, we are using only a limited portion of this range. Therefore, the image would appear dark gray with very little contrast.

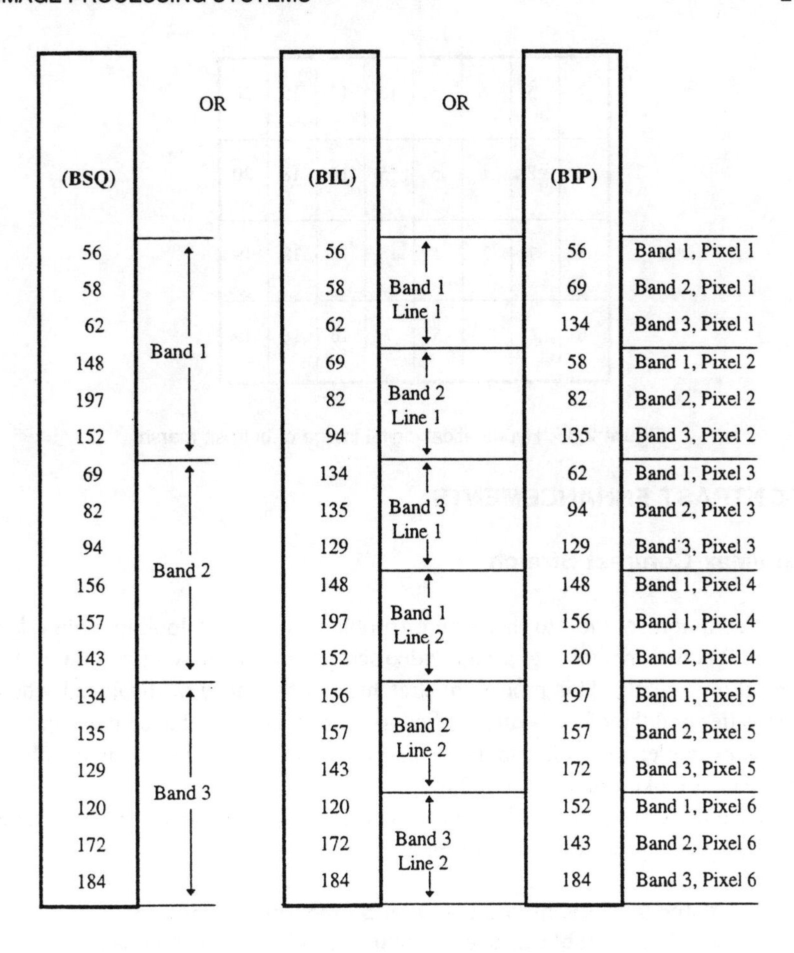

Figure 2.6. Common data-packaging formats for storage of image data (BSQ = Band Sequential, BIL = Band Interleaved by Line, BIP = Band Interleaved by Pixel).

Digital Image

245	5	243
243	6	242
200	15	213

Displayed Image

BRIGHT	DARK	BRIGHT
BRIGHT	DARK	BRIGHT
BRIGHT	DARK	BRIGHT

Figure 2.7. Simplified example of single-band image display.

6	5	6	5	15	17	19	20
6	8	6	5	15	16	18	20
6	7	7	5	15	16	18	19
6	7	7	5	7	16	18	19

Figure 2.8. Hypothetical digital image of bulrush marsh.

CONTRAST ENHANCEMENTS

Min/Max Contrast Stretch

A common solution to this type of problem is to use a lookup table (also called function memory) to assign video screen intensities as a function of the digital image data. This procedure stretches out the range of displayed video intensities and therefore is often called a contrast stretch enhancement. In this marsh example, we could use the minimum and maximum image pixel values to create a lookup table (Table 2.2).

Since there are 255 possible video intensities, we can create a lookup table values by using the following formula:

Table 2.2. Example of a function memory or lookup table to scale monitor pixel video intensities

Pixel Digital Value	Lookup Value	Monitor Video Intensity
6	0	0
7	18	18
8	36	36
9	55	55
10	73	73
11	91	91
12	109	109
13	128	128
14	146	146
15	164	164
16	182	182
17	200	200
18	217	217
19	237	237
20	255	255

$$\text{Screen Pixel Intensity} = \frac{(\text{Pixel Digital Value} - \text{Minimum Pixel Digital Value})}{(\text{Max. Pixel Digital Value} - \text{Min. Pixel Digital Value})} \times 255$$

For example, the digital value for pixel 1,1 (row 1, col 1) in the marsh image was 15. The function memory (or lookup value) for 15 is:

$$\frac{(15-6)}{(20-6)} \times 255 = 163.9$$

Since the screen intensity values are integers from 0 to 255, the lookup values are rounded to the nearest integer. Therefore 163.9 is rounded to 164 and any pixel with a digital value of 15 is displayed with a screen intensity of 164.

This approach of using minimum and maximums for creating a lookup table is sometimes not a good strategy. Why?

We might have a satellite image where there are millions of pixels and therefore the minimum and maximum values may represent rare features on the image. For example, we might be interested in displaying loblolly pine sapling plantations with a SPOT panchromatic image. Let's assume the pine plantations have digital values ranging from 14 to 30. The histogram of digital values from the entire image might look like the following:

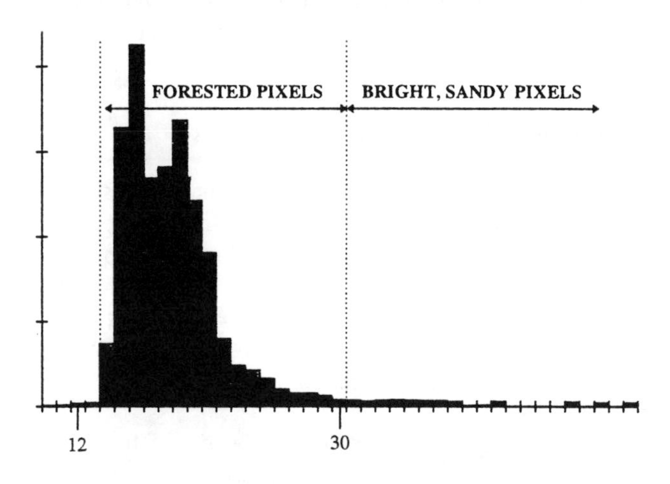

Figure 2.9. Histogram of pixel digital values.

Due to sandy roads, we have a few bright pixels on the image (digital values greater than 29). If we created a lookup table based on the minimum digital value of 12 and a maximum digital value of 50, we would get good overall contrast and the sandy roads would be evident on the image. However, since our goal is to inventory sapling plantations range in digital values from 14 to

30, a lookup table based on these values would be provide a better visual contrast between the plantation areas and the surrounding forest.

Linear Stretch Based on Standard Deviations

Image processing programs sometimes use the following default values for lookup tables:

lower limit = mean – (2 × standard deviation of image digital values)

upper limit = mean + (2 × standard deviation of image digital values)

There is nothing magical about these default limits. If the image digital values have a bell-shaped (normal) distribution, then these values are assumed to represent the range covering 95% of the image digital values. Therefore, this enhancement is sometimes referred to as a 95% contrast stretch.

Sometimes there are more appropriate upper and lower limits. For example, a waterfowl biologist may be interested in seeing the distribution of sago pondweed and therefore would want to set a lookup table to enhance the contrast within a range of low digital values (areas that are predominantly water). On the other hand, a forest hydrologist may be interested in seeing high contrast within snow-covered areas and therefore may use a lookup table with a much higher lower limit. In these examples, the user would want to set the upper and lower limits of the linear contrast stretch, and not accept the default 95% contrast stretch values.

Histogram Equalization

Another common strategy in image enhancement is to assign video intensities so that approximately the same number of image pixels are displayed with each video intensity. This approach is called histogram equalization. Let's work through an example. Imagine that you have a printer that can print in 8 shades of grey (0 = black, 1 = very dark gray, ..., 6 = very light gray, 7 = white). The image ranges in pixel values from 50 to 70 as follows. We first apply a min/max linear contrast stretch to print out the image:

$$\text{Video Intensity} = \left[\frac{(\text{Pixel Digital Value} - 50)}{(70 - 50)}\right] * 255$$

The goal of histogram equalization is to "flatten out" the histogram such that there is a more even distribution of number of pixels assigned to each video intensity (or shade class if working with printer shades). Because print shades or video intensities are discrete values, histogram equalization will usually not create a perfectly uniform display histogram, but it will create a more uniform display histogram relative to the original histogram.

Table 2.3. Print shade values assigned using min/max linear contrast stretch

Pixel Value	Frequency	Print Shade Assigned by Min/Max Stretch
50	1	0
51	2	0
52	1	1
53	2	1
54	2	1
55	6	2
56	8	2
57	5	2
58	4	3
59	2	3
60	1	3
61	1	4
62	0	4
63	1	5
64	2	5
65	1	5
66	2	6
67	4	6
68	7	6
69	8	7
70	4	7

Using our previous example, if we had a perfectly even histogram, how many pixels would be assigned to each shade class? Since we have 64 pixels and 8 shade classes (0 through 7) our ideal equalized histogram would be perfectly even with 8 pixels in each shade class.

We can perform the histogram equalization using a spreadsheet approach (Table 2.4). Our original image had 21 values ranging from 50 to 70. If these

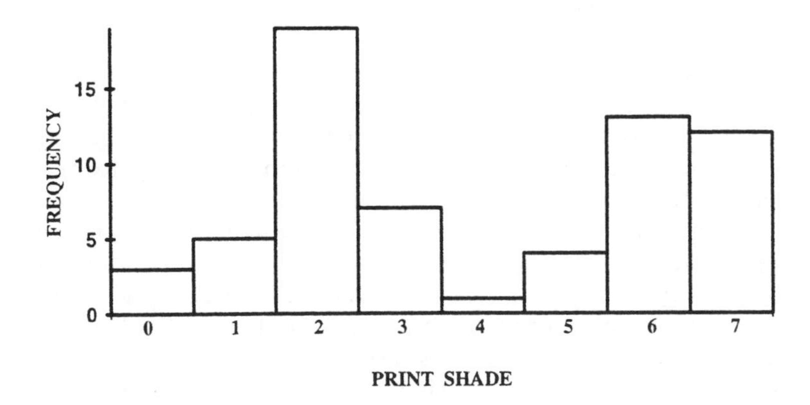

Figure 2.10. Histogram of print shades assigned using min/max linear contrast stretch.

Table 2.4. New print shade values assigned by histogram equalization

Pixel Value	Number of Pixels	Cumulative Actual	Cumulative Target	Old Print Shade	New Print Shade
50	1	1	3.05	0	0
51	2	3	6.10	0	0
52	1	4	9.15	1	0
53	2	6	12.20	1	0
54	2	8	15.25	1	1
55	6	14	18.30	2	1
56	8	22	21.35	2	2
57	5	27	24.40	2	3
58	4	31	27.45	3	3
59	2	33	30.50	3	3
60	1	34	33.55	3	4
61	1	35	36.60	4	4
62	0	35	39.65	4	4
63	1	36	42.70	5	4
64	2	38	45.75	5	4
65	1	39	48.80	5	4
66	2	41	51.85	6	5
67	4	45	54.90	6	5
68	7	52	57.95	6	6
69	8	60	61.00	7	7
70	4	64	64.05	7	7

values were equally distributed, each value would have a frequency of slightly more than 3 pixels (64 total image pixels / 21 values = 3.05). The first step is to fill out this target frequency as a cumulative (or sum) frequency for pixel values ranging from 50 to 70. The next step is to fill out a column showing the actual cumulative frequency values for the image. We already have an original print shade assigned for each pixel value. The final step is to assign new print shades by comparing the actual cumulative frequency values with the target cumulative frequency values. Assign the new print shade value by finding the cumulative target value that is closest, but not less than, the cumulative actual value. The new print shade will be the old print shade value from this row. For example, pixel values 52 has a cumulative actual value of 4. The closest cumulative target value is 6.10 (old print shade 0). Therefore, pixel value 52 will be assigned 0 as the new print shade value. Pixel value 53 with a cumulative actual value of 6 will also be assigned 0 as the new print shade value, because the closest cumulative target is 6.10 (old print shade 0). Pixel value 54 will be assigned a new print shade of 1, because the closest cumulative target to 8 is 9.15 (old print shade 1).

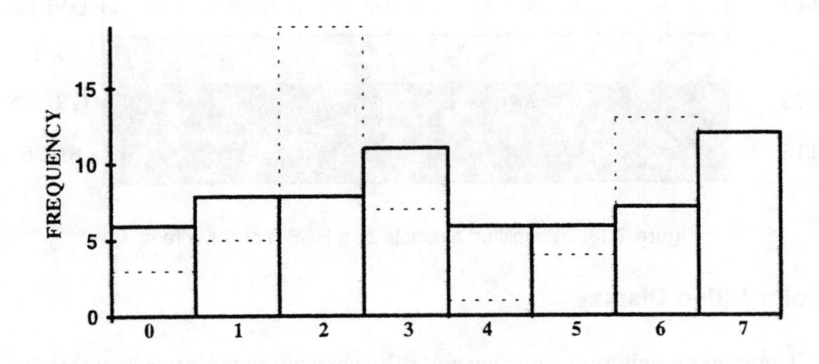

Figure 2.11. Histogram of print shade assignments after histogram equalization.

DISPLAY OF COLOR IMAGES

We can create any color of light by combining three additive primary colors: red, green, and blue *light*. For example, white light is composed of equal proportions of bright red, bright green, and bright blue light. Black is the absence of all primary colors of light. Gray color can be produced by mixing equal proportions of lower brightness red, green, and blue lights. The following figure illustrates how the three additive primary colors can be used to produce other colors.

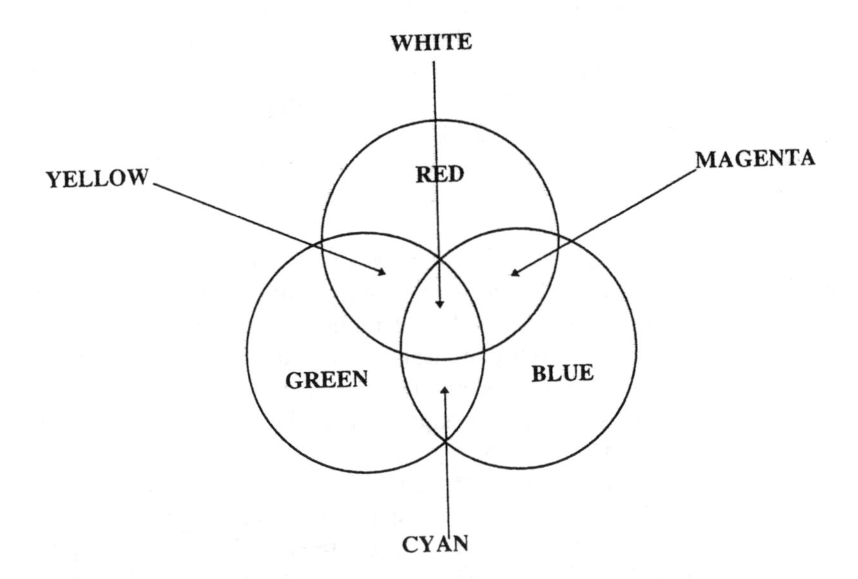

Figure 2.12. Colors produced by combining the additive primary colors (red, green, and blue).

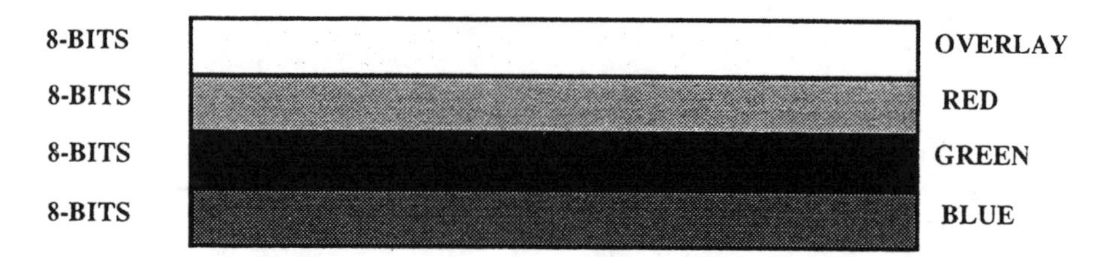

Figure 2.13. Simplified example of a RGB monitor screen.

Color Video Display

Computer monitors can display colors by controlling the intensity or brightness of red, green, and blue phosphors. These monitors are often called RGB (red, green, blue) monitors. Imagine an RGB monitor that consists of a sandwich of three computer screens: a red screen, a green screen, and a blue screen plane (Figure 2.13).

Each plane consists of a fixed number of rows and columns as determined by an image processing board. Some common examples are 512 rows by 512 columns and 1024 rows by 1024 columns. The color on the computer monitor is controlled by varying the intensity or brightness of the red, green, and blue planes at each screen pixel. Typically, the brightness values can be varied from 0 to 255 (8 bits). This means the image processing board must have 8 bits of memory for each RGB plane or 24 total bits. Image processing boards are often referred to by their memory capacity. For example, a 512 × 512 × 24 bit board would consist of memory for 512 rows by 512 columns by 24 bits (8 bits for the red plane, 8 bits for the green plane, and 8 bits for the blue plane).

Let's take a very simple example. Imagine that we have an RGB monitor and image processing card that display 3 rows by 5 columns and have the video intensity values shown in Figure 2.14.

Try to figure out what colors you will see for each cell on the computer monitor screen. Then turn to the next page (Figure 2.15).

If we have a 24 bit display device — 8 bits in each red, green, and blue plane (8 bits allow for a video intensity range of 0 to 255) — then how many different colors can we display? 2^{24} = 16.7 million different colors! Of course we can not visually distinguish all 16.7 million colors; however, such a system does allow us the ability to display subtle differences in colors that may be important in the visual interpretation of images for natural resource applications including vegetation mapping, plant vigor assessment, and water quality assessment.

Sometimes, you might want to display features on top of a color digital image. For example you might want to display the boundaries of a wilderness area, the location of a riparian buffer, range allotment boundaries, logging roads, or property lines on a digital image. To display such features, many image processing systems use an overlay plane (also called graphics or annotation plane) that can be thought of as a fourth layer on top of our red, green,

RED PLANE

255	0	0	255	0
100	0	0	200	100
255	0	255	125	95

GREEN PLANE

0	255	0	255	255
0	100	0	200	100
255	0	150	255	55

BLUE PLANE

0	0	255	0	255
0	0	100	200	100
255	0	0	0	0

Figure 2.14. Color image produced by using digital values to control red, green, and blue video intensities.

RGB MONITOR SCREEN

BRIGHT RED	BRIGHT GREEN	BRIGHT BLUE	BRIGHT YELLOW	BRIGHT CYAN
DARK RED	DARK GREEN	DARK BLUE	LIGHT GRAY	DARKER GRAY
WHITE	BLACK	ORANGE	YELLOW-- GREEN	BROWN

Figure 2.15. Resulting color display on a RGB computer monitor screen.

and blue sandwich. The graphics plane can display different intensities depending upon how many bits of memory the plane has. For example, a 8-bit overlay plane would be capable of displaying intensities ranging from 0 to 255.

Color Image Printing

Color printing is analogous to color video display except that three colored dyes are used. Cartographers call these dyes process colors. These colors are also called subtractive primaries (cyan, magenta, and yellow) because they subtract (or absorb) one of the three primary colors. For example, a yellow dye absorbs blue light and reflects the remaining primary colors of red and green (remember that equal proportions of red and green light produce yellow light). Magenta dye absorbs green light and reflects equal proportions of red and blue light (therefore magenta is a reddish-blue color). Cyan dye absorbs red light and reflects equal proportions of green and blue light (therefore cyan appears as a greenish-blue color). Various colors can be produced by combining various proportions of cyan, magenta, and yellow ink on paper. To produce the color black, we could soak up a spot on the paper with cyan (minus red), magenta (minus green) and yellow (minus blue) inks. However, it is simpler, less expensive, and neater, to use black ink instead of soaking up a location on the paper with three inks. Therefore most color printers have black ink as well as the three subtractive primary colors (cyan, magenta, yellow).

You can draw an aid to remembering the subtractive and additive primary colors using Fahsi's triangle system (Figure 2.16). Draw a triangle and label corner points (starting at the top of the triangle) in clockwise order with the additive primary colors in alphabetical order: blue, green, red. Then label the legs of the triangle (starting with the right leg) in counterclockwise order with the subtractive primaries in alphabetical order: cyan, magenta, yellow. The subtractive primary will be opposite of its additive primary. For ex-

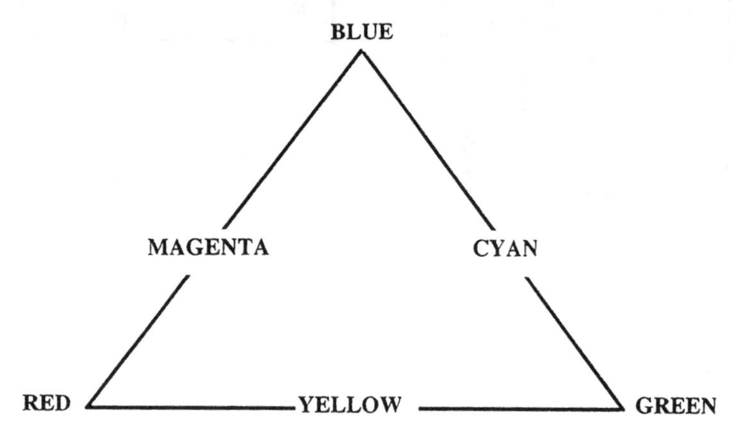

Figure 2.16. Fahsi's triangle: a tool to help remember subtractive and additive primary colors.

ample, yellow dye reflects equal combinations of red and green and subtracts out (or absorbs) blue light.

IMAGE MAGNIFICATION AND REDUCTION

A computer monitor can only display a limited number of image rows and columns. Since digital images can have thousands of rows and columns, every image pixel cannot be simultaneously displayed on the monitor screen. Therefore, an image is displayed at a certain reduction or magnification factor. For example, imagine that we have the following single-band image file:

DATA FILE:

	COL #1	COL #2	COL #3	COL #4	COL #5	COL #6	COL #7	COL #8	COL #9	COL #10
ROW#1	57	64	67	69	69	72	68	70	69	71
ROW#2	55	63	65	67	68	79	69	70	71	72
ROW#3	58	64	49	37	32	30	31	32	35	55
ROW#4	56	63	39	37	34	33	33	33	31	40
ROW#5	57	63	59	55	54	53	50	49	47	54
ROW#6	57	64	67	69	69	72	68	70	69	71
ROW#7	55	63	65	67	68	70	69	70	71	72
ROW#8	58	64	40	37	32	70	91	92	35	55
ROW#9	56	63	39	37	34	73	83	93	31	40
ROW#10	57	63	59	55	54	53	50	49	47	54

Figure 2.17. Example of a 10 column by 10 row single-band image file.

Now assume an extreme example in which we have a monitor that only can display 5 rows and 5 columns. We can display the entire image if we display it with a reduction factor of 2 and centered at file coordinates 5,5 (Figure 2.18).

COMPUTER MONITOR SCREEN

57	67	69	68	69
58	49	32	31	35
57	59	54	50	47
55	65	68	69	71
56	39	34	83	31

Figure 2.18. Example of single band image displayed with reduction factor of 2.

COMPUTER MONITOR SCREEN

49	37	32	30	31
39	37	34	33	33
59	55	54	53	50
67	69	69	72	68
65	67	68	70	69

Figure 2.19. Example of single band image displayed with a magnification factor of 1X.

Or we can display the image with a magnification factor of 1, centered at file coordinates 5,5 (Figure 2.19).

CHAPTER 2 PROBLEMS

1) Landsat MSS data are recorded in 6-bit format. What is the possible range of digital numbers using this format?

   When MSS data are processed on earth, they are stretched from 6-bit to 7-bit format. What is the possible range of digital numbers in 7-bit format?

2) Given the following data on 9-track tape, fill in the proper values for each pixel in the image:

IMAGE:

BAND 1:

BAND 2:

BAND 3:

BAND 1:

BAND 2:

BAND 3:

9-track tape information:
Data packaged as BIL (band interleaved by line)
Number of rows: 2 Number of columns: 2 Number of bands: 3

Tape File:

Parity Bit	Bit#8 (2^7)	Bit#7 (2^6)	Bit#6 (2^5)	Bit#5 (2^4)	Bit#4 (2^3)	Bit#3 (2^2)	Bit#2 (2^1)	Bit#1 (2^0)
	0	0	0	1	0	1	0	1
	0	0	0	1	1	0	1	0
	0	0	1	0	0	0	0	0
	0	0	0	1	1	1	0	0
	0	0	1	0	1	1	1	1
	0	0	1	1	1	0	0	1
	0	0	1	1	1	0	1	1
	0	0	1	1	0	1	1	0
	0	1	0	0	0	0	1	1
	0	1	0	1	0	0	0	0
	0	0	1	0	1	1	1	1
	0	0	1	1	1	0	0	1

3) Imagine you have a single band image that contains the following digital numbers (DNs):

121	109	132	115	110
113	112	115	123	122

Compute the mean and standard deviation of these digital values.
Mean = Standard Deviation =
Create a lookup table with following limits:
lower limit = mean – 2 * standard deviation
upper limit = mean + 2 * standard deviation

4) Match the following colors with the appropriate RGB video intensities:

COLOR CHOICES:	DISPLAYED COLOR:	RED VIDEO INTENSITY	GREEN VIDEO INTENSITY	BLUE VIDEO INTENSITY
1) Red	______	255	255	255
2) Green	______	125	125	125
3) Blue	______	0	0	0
4) White	______	255	0	0
5) Black	______	255	150	0
6) Gray	______	0	255	0
7) Cyan	______	0	0	255
8) Magenta	______	255	255	0
9) Yellow	______	0	0	70
10) Brown	______	65	255	130
11) Sand	______	150	150	0
12) Orange	______	0	255	255

13) Navy Blue ______	255	0	255
14) Mint Green ______	215	167	116
15) Avocado ______	95	60	0

5) **Given the following digital image file, assume you have a computer monitor that can display only 3 rows by 3 columns.**

	COL #1	COL #2	COL #3	COL #4	COL #5	COL #6	COL #7	COL #8	COL #9	COL #10
ROW#1	57	54	67	69	72	68	70	69	71	71
ROW#2	55	63	65	68	70	70	72	71	69	67
ROW#3	58	64	40	32	30	32	33	36	38	39
ROW#4	56	63	39	34	33	36	37	39	40	43
ROW#5	57	63	59	54	53	54	55	62	63	65
ROW#6	55	64	67	69	72	70	70	70	69	69
ROW#7	57	63	65	67	70	67	69	61	67	68
ROW#8	58	64	40	40	70	69	65	63	65	70
ROW#9	56	64	39	37	73	71	69	65	69	72
ROW#10	57	63	59	55	53	72	70	68	72	75

Fill in the appropriate monitor pixel digital values, assuming this image was displayed with a reduction factor of 2 and centered at file coordinates 4,4:

MONITOR SCREEN:

6) You have a computer monitor that displays red, green, and blue in 0–255 video intensities. You also have a color ink-jet printer that prints with cyan, magenta, and yellow inks in 0–15 shades. Write three functions that will convert any red, green, blue video intensities to the proper cyan, magenta, and yellow print shades such that a color image printed out will appear similar to the color image displayed on the computer monitor screen.

Cyan =
Magenta =
Yellow =

7) Read in the following 9-track tape given the following information:

packaging = bil nrows = 3 ncols = 2 nbands = 3

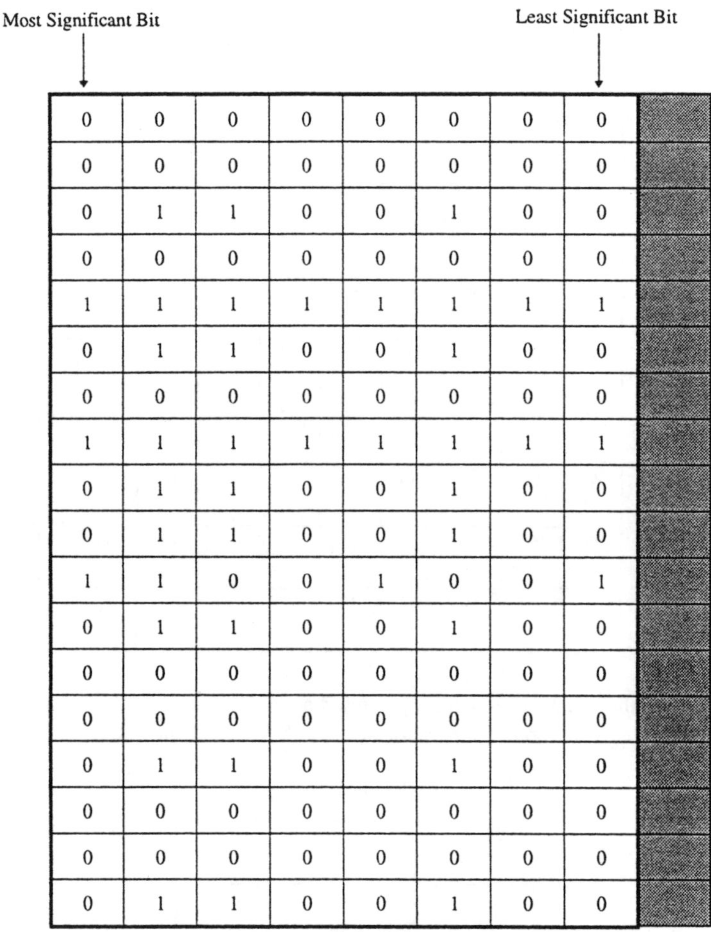

0	0	0	0	0	0	0	0
0	0	0	0	0	0	0	0
0	1	1	0	0	1	0	0
0	0	0	0	0	0	0	0
1	1	1	1	1	1	1	1
0	1	1	0	0	1	0	0
0	0	0	0	0	0	0	0
1	1	1	1	1	1	1	1
0	1	1	0	0	1	0	0
0	1	1	0	0	1	0	0
1	1	0	0	1	0	0	1
0	1	1	0	0	1	0	0
0	0	0	0	0	0	0	0
0	0	0	0	0	0	0	0
0	1	1	0	0	1	0	0
0	0	0	0	0	0	0	0
0	0	0	0	0	0	0	0
0	1	1	0	0	1	0	0

What will be the pixel values for the corresponding image?

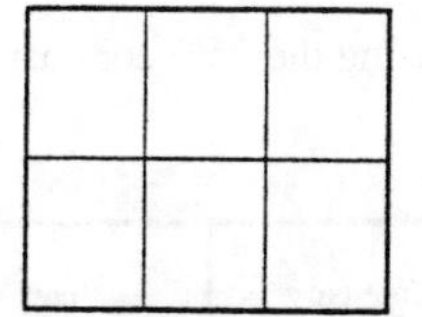

If you assigned band 3 as a red overlay, band 2 as a green overlay, and band 1 as a blue overlay, what colors will the pixels be displayed as?

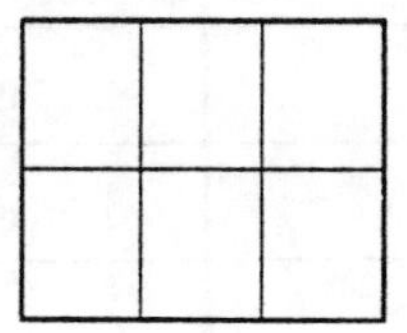

If you assigned band 2 to the red, green, and blue overlay, what colors will the pixels be displayed as?

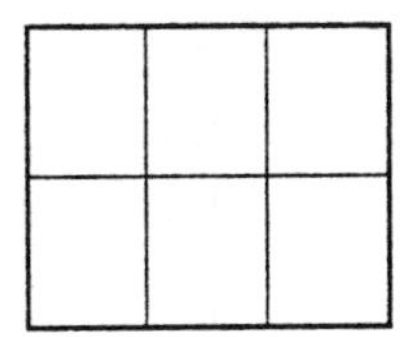

8) What term does the following 3 bytes represent using ASCII coding?

0	1	0	0	1	0	0	1

0	1	0	0	1	0	1	1

0	1	0	1	0	0	1	1

9) You have the following digital image:

5	8	14	14	14	10	9	7	7	8
5	7	14	13	13	10	9	6	7	7
6	7	11	12	13	11	10	6	6	6
9	8	9	11	12	10	10	6	5	6
17	18	20	15	14	13	10	6	5	6
19	20	22	20	19	19	12	6	6	6
26	27	17	19	19	19	14	8	9	9
29	28	16	20	19	19	17	10	10	9
32	30	21	21	20	21	17	11	10	10
31	30	26	25	22	22	15	12	10	10

Develop a lookup table using the 95% contrast stretch enhancement for this image.

Pixel Value	Video Intensity	Pixel Value	Video Intensity
0		17	
1		18	
2		19	
3		20	
4		21	
5		22	
6		23	
7		24	
8		25	
9		26	
10		27	
11		28	
12		29	
13		30	
14		31	
15		32	
16		33	

10) Using the 10-row by 10-column image from problem 9, develop a min/max linear stretch. What are the pixel video intensity values after the linear stretch enhancement?

11) Develop a histogram equalization from the min/max linear stretched image in problem 10. Fill in the appropriate video intensity values after the histogram equalization enhancement:

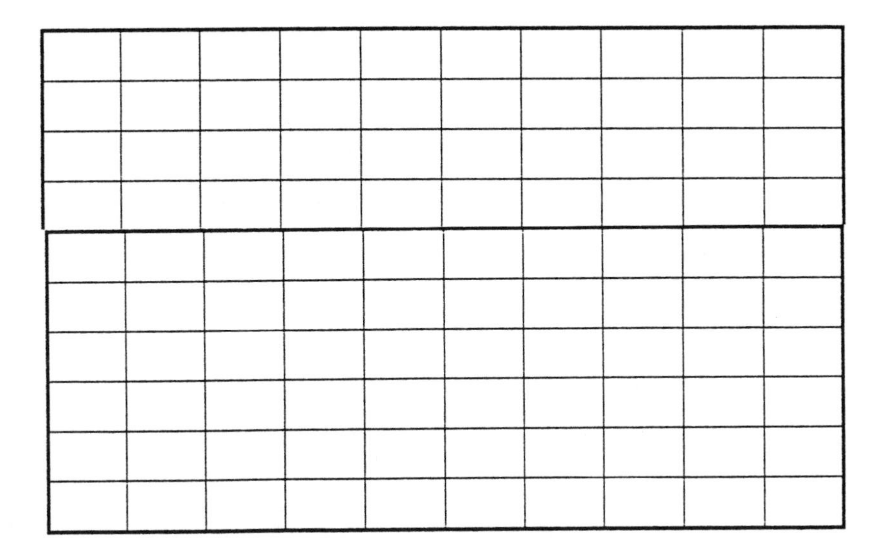

ADDITIONAL READINGS

Aranuvachapun, S. and D. E. Walling. 1987. The use of a microcomputer for image analysis. *International Journal of Remote Sensing.* 8:1385–1397.

Cooke, D. F. 1987. Map storage on CD ROM. *Byte Magazine.* 12(8):129–138.

Di, L and D. C. Rundquist. 1988. Color-*composite image generation on an eight-bit graphics workstation. Photogrammetric Engineering and Remote Sensing.* 54:1745–1748.

Di, L. and D. C. Rundquist. 1991. A classification-based method for color-composite image generation on an eight-bit graphics workstation. *Geocarto International.* 6:23–29.

Heckbert, P. 1982. Color image quantization for frame buffer display. *Computer Graphics.* 16:297–304.

Kiefer, R. W. and F. J. Gunther. Digital image processing using the Apple-II microcomputer. *Photogrammetric Engineering and Remote Sensing.* 49:1167–1174.

Mather, P. M. 1991. *Computer Applications in Geography.* J. Wiley & Sons Inc., New York. 257 pp.

Welch, R. A. 1983. Microcomputers in the mapping sciences. *Computer Graphics World.* 6:33–42.

CHAPTER 3

Spectral Regions

INTRODUCTION

Satellite images are digital images recorded from specific spectral regions. The main objective of this chapter is to show why spectral characteristics of natural cover types vary, and how these spectral patterns can be utilized for satellite remote sensing of natural resources.

A crucial assumption in satellite remote sensing is that cover types are spectrally separable. For example, water, healthy vegetation, and dry soil all have different characteristic reflectance patterns:

Table 3.1. Example of typical reflectance patterns for water, soil, and a maple forest

Cover Type	Near-Infrared Reflectance	Mid-Infrared Reflectance
Water	Low	Low
Dry Soil	Medium	High
Maple Forest	High	Medium

A digital image using near and mid-infrared sensor detectors might have the following digital values (Figure 3.1). Assuming the range of digital values from 0 to 50 is "low," 51 to 100 is "medium," and greater than 100 represents "high" reflectance, predict the cover type (water, maple forest, or dry soil) for each pixel by filling in the appropriate cover type:

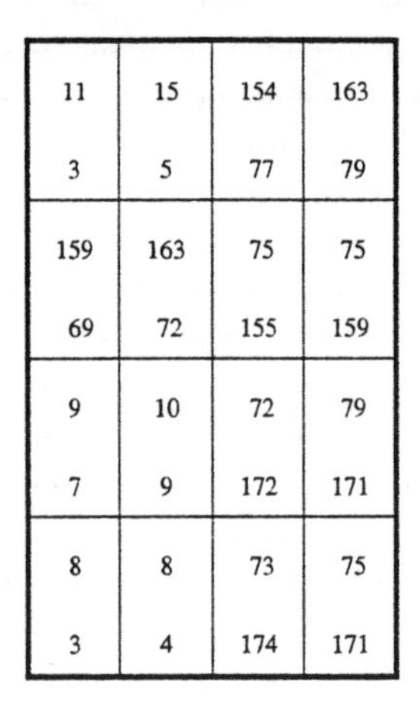

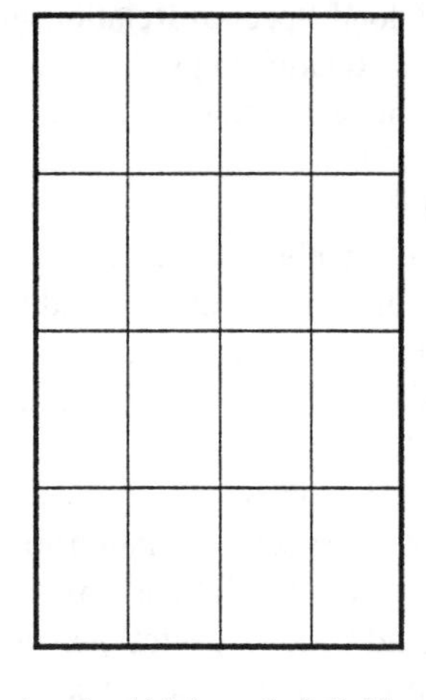

Figure 3.1. Two-band (near-infrared and mid-infrared) digital image.

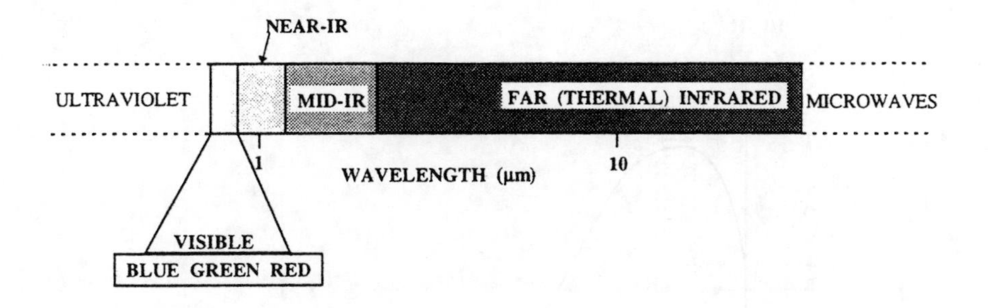

Figure 3.2. Spectral regions along the electromagnetic spectrum.

SPECTRAL REGIONS

Electromagnetic Spectrum

Remote sensing systems detect radiation from various portions of a continuous spectrum of radiation, called the electromagnetic spectrum (Figure 3.2). For example, our eyes as a remote sensing system can detect reflectance from a very narrow spectral region called the visible region of the electromagnetic spectrum. You will learn that other invisible spectral regions are very important for some natural resource applications, and therefore, SPOT HRV, AVHRR, and Landsat sensors have spectral bands from invisible regions.

Spectral Distribution of Solar Energy

Why do most animals see light in the wavelength range of 0.4 to 0.7 μm? From Figure 3.3 you can see that the peak of solar energy occurs in this visible wavelength region. As we move from short wavelengths to longer wavelengths, the energy content decreases. The lower energy content of long wavelength radiation generally means that ground-pixel size must be large in order to receive enough detectable energy from long wavelengths. This is why, for example, Landsat Thematic Mapper has 120-meter pixels in the long-wavelength thermal infrared spectral band and 30-meter pixels in the shorter-wavelength spectral bands. This is also why Landsat TM band widths increase with increasing wavelengths (Figure 3.4).

Terminology

Energy, the capacity to do work, is expressed in joules (J). The flow of energy (for example, solar energy from the sun to the earth) is termed *flux* and is expressed in watts (W). Density implies distribution per unit area or volume. Therefore, the term *radiant flux density* or *irradiance* is the rate of incident energy per unit area and is often expressed in watts per square meter. *Radiance* denotes the irradiance as viewed through a three-dimensional angle. This solid

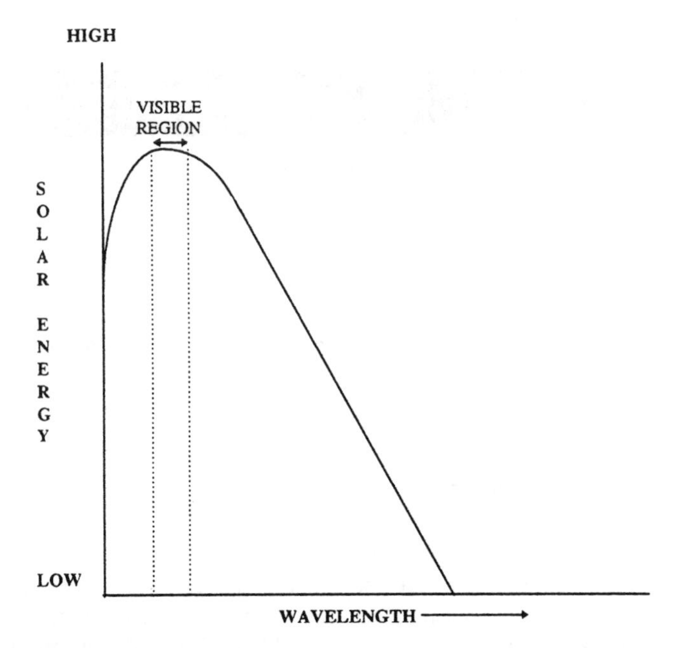

Figure 3.3. Spectral distribution of energy radiated from the sun at 6000° K.

angle is usually expressed in steradians, and radiance is often expressed in watts per square meter per steradian. The term *spectral* is often used as a prefix to denote a specific range of wavelengths. For example, *spectral* radiance for a given wavelength region would be expressed in watts per square meter per steradian per unit wavelength.

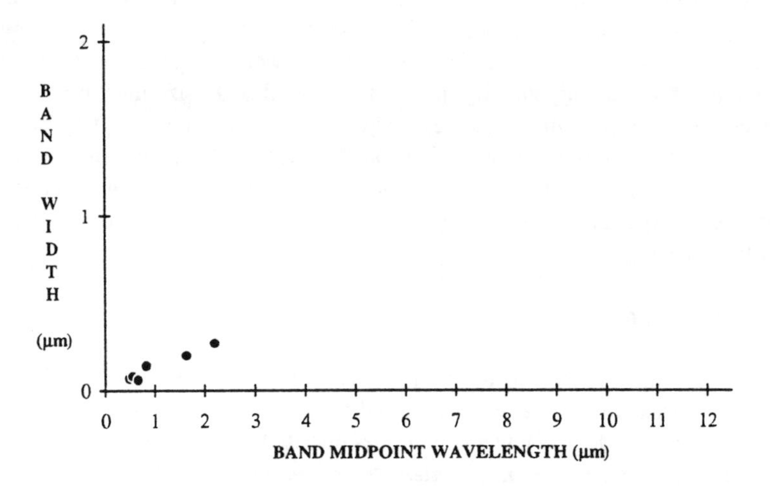

Figure 3.4. Increase in width of Landsat Thematic Mapper bands as a function of spectral wavelength.

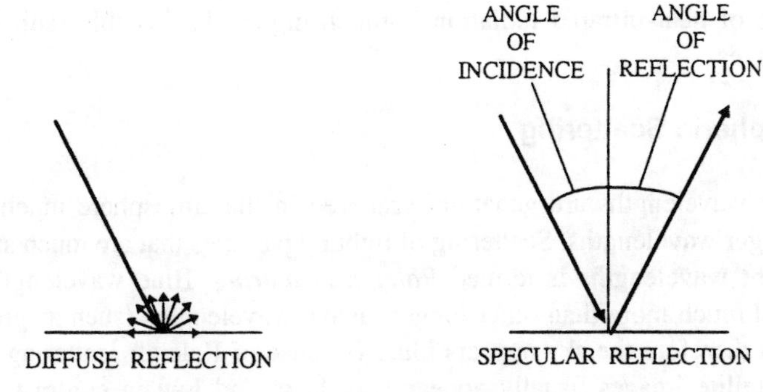

Figure 3.5. Diffuse versus specular reflectance.

Insolation-Earth Surface Interactions

*In*coming *sol*ar radi*ation* is often called *insolation.* About half of the solar radiation entering the atmosphere reaches the earth's surface and half is reflected or absorbed within the earth's atmoshpere. Insolation reaching the earth's surface is either reflected, absorbed, or transmitted.

Reflection

Reflection occurs when incoming radiation is redirected by a non-transparent surface. If the surface is rough, incoming radiation is scattered nearly equally in all directions. Such a surface is called a Lambertian surface or a perfectly diffuse reflector. A uniform lawn would approximate a *Lambertian* surface for visible wavelengths. If the surface is smooth, incoming radiation will be reflected at a angle equal to the angle of incidence (Figure 3.5). Such a surface is called a *specular* reflector. A mirror would approximate a specular reflector for visible wavelengths. At low sun angles, such as in high latitudes during the winter, vegetated surfaces tend towards specular reflection.

Reflectance or *reflection factor* refers to the ratio of total reflected radiation divided by the total incoming radiation. If the reflectance is observed for a fixed viewing angle, then the term *bi-directional* reflectance is used. *Spectral reflectance* is the reflectance ratio for a given range of wavelengths. *Albedo* refers to the reflectance ratio for total incoming solar radiation (wavelengths ranging from 0.3 to 4.0 μm).

Absorption and Transmission

Insolation that is not reflected by a surface, is either absorbed or transmitted. Absorption and transmittance both vary with wavelength. For example, chlorophyll absorbs radiation primarily from the violet-blue and red regions of the visible spectrum and reflects radiation from the green region. Likewise, transmittance of near-infrared radiation is much higher than visible radiation in most leaves.

Atmospheric Scattering

Short wavelengths are generally scattered in the atmosphere much more than longer wavelengths. Scattering of light by particles that are much smaller than light wavelengths is termed *Raleigh scattering*. Blue wavelengths are scattered much more than other longer visible wavelengths (such as green or red) and therefore the sky appears blue. Because of Raleigh scattering, blue-band satellite images usually appear very hazy and low in contrast, while images using longer wavelength bands such as near-infrared will often appear relatively clear with high contrast. You will learn about corrections that can be applied for atmospheric scattering effects in the next chapter.

Spectral Bands

It is important to understand the characteristic spectral pattern of various natural features, especially in the spectral regions commonly used by satellite sensors for natural resources applications (Table 3.2).

VEGETATION SPECTRAL RELATIONSHIPS

Visible Spectral Region

Vegetation generally has low reflectance and low transmittance in the visible part of the spectrum. This is mainly due to plant pigments absorbing visible light. For example, chlorophyll pigments absorb violet-blue and red light for photosynthic energy. Green light is not absorbed for photosynthesis and therefore most plants appear green (Figure 3.6).

In the autumn, some plant leaves such as foliage of aspen, cottonwood, and tuliptree turn from green to a brilliant yellow. This change in foliage color is caused by the normal autumn breakdown of chlorophyll (which usually is the dominant pigment during the summer). After the breakdown of chlorophyll, other pigments such as carotenes and xanthophylls become dominant and therefore the foliage color changes from green to yellow. Carotene and xanthophyll pigments absorb blue light and reflect green and red light (remember, yellow light is the equal combination of green and red light).

In the autumn, foliage from plants such as red maple, sumac, and huckleberry often change from green to bright red. These plants produce large quantities of the pigment anthocyanin (which absorbs blue and green light, and reflects red light) which creates the red fall foliage on these plants (Figure 3.7).

Table 3.2. Landsat, SPOT, and AVHRR spectral bands.

Landsat Multispectral Scanner (56 by 79 m pixels)	Landsat Thermatic Mapper (30 by 30 m pixels)	SPOT HRV Multispectral Mode (20 by 20 m pixels)	SPOT HRV Panchromatic Mode (10 by 10 m pixels)	AVHRR (1 by 1 km pixels)
	Band 1: 0.45–0.52 μm (Blue Band)			
Band 1: 0.5–0.6 μm (Green Band)	Band 2: 0.52–0.60 μm (Green Band)	Band 1: 0.50–0.59 μm (Green Band)	Band 1: 0.51–0.73 μm	
Band 2: 0.6–0.7 μm (Red Band)	Band 3: 0.63–0.69 μm (Red Band)	Band 2: 0.61–0.68 μm (Red Band)	(Green to Near-IR)	Band 1: 0.58–0.68 μm (Red Band)
Band 3: 0.7–0.8 μm (Near-IR Band #1)	Band 4: 0.76–0.90 μm (Near-IR Band)	Band 3: 0.79–0.89 μm (Near-IR Band)		Band 2: 0.73–1.10 μm (Near-IR Band)
Band 4: 0.8–1.1 μm (Near-IR Band #2)	Band 5: 1.55–1.75 μm (Mid-IR Band)			Band 3: (3.55–3.93) Thermal #1
	Band 6[a]: 10.4–12.5 μm (Thermal Band)			Band 4: (10.30–11.30) Thermal #2
	Band 7[b]: 2.08–2.35 μm (Mid-IR Band)			Band 5: (11.50–12.50) Thermal #3

[a] Band 6 (Thermal Infrared) has 120 meter square ground pixels.

[b] Band 7 (Mid-infrared #2) was a late addition in the planning Landsat Thematic Mapper and consequently is sequenced after the thermal infrared band.

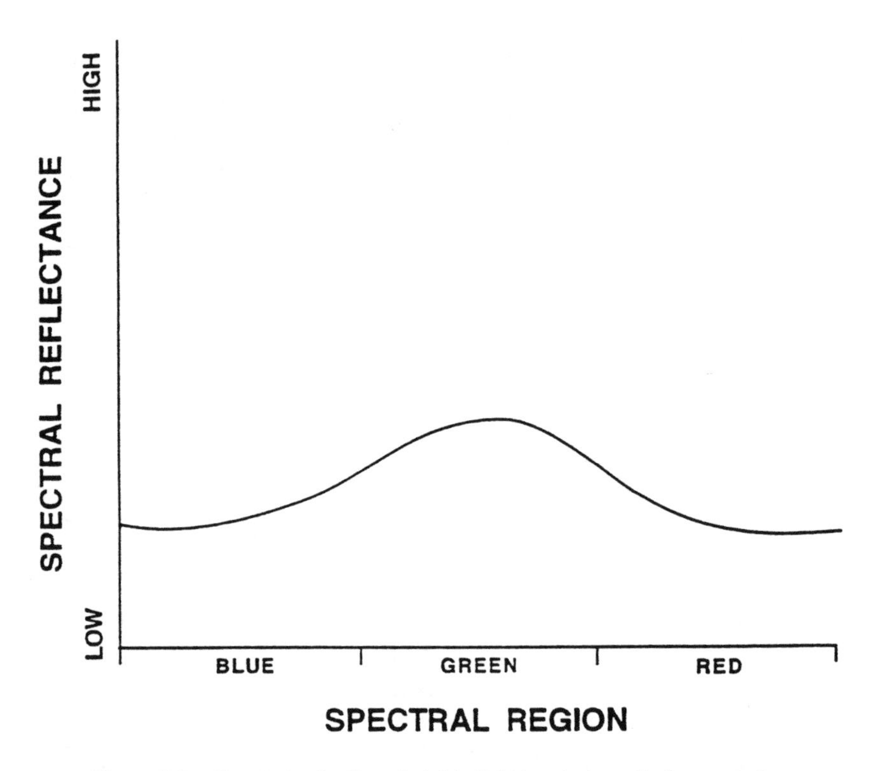

Figure 3.6. Spectral reflection of visible light by photosynthetic vegetation.

Near-Infrared Spectral Region

Plants generally reflect radiation highly in the near infrared region. This is mainly due to the high air/cell interface area within leaves (Gates 1970, Gausman et al. 1977). This can be shown by infiltrating leaves with water under vacuum. As the water fills the air gaps, there is a decrease of multiple scattering and a decrease in near-infrared reflectance (Figure 3.8).

The high near-infrared reflectance of plants sometimes is erroneously attributed to high chlorophyll content of leaves. We can show that chlorophyll content of leaves does not affect the reflectance of near-infrared radiation by examining the reflectance spectra of a green and white geranium leaf. The white portion of the leaf, which lacks chlorophyll, will have nearly the same near-infrared reflectance as the green portion of the leaf (Figure 3.9). Both the white (nonchlorophyll) and green (chlorophyll) portions of the leaf will have high near-infrared reflectance due to a high air-cell interface area.

Plants with different internal leaf cell structure will often vary greatly in near infrared reflectance. The reflectance of near-infrared radiation will also vary depending upon the shape and orientation of plant leaves (Williams 1991). Therefore, near-infrared reflectance values are often more useful than visible reflectance values in distinguishing forest types such as aspen versus spruce (Figure 3.10).

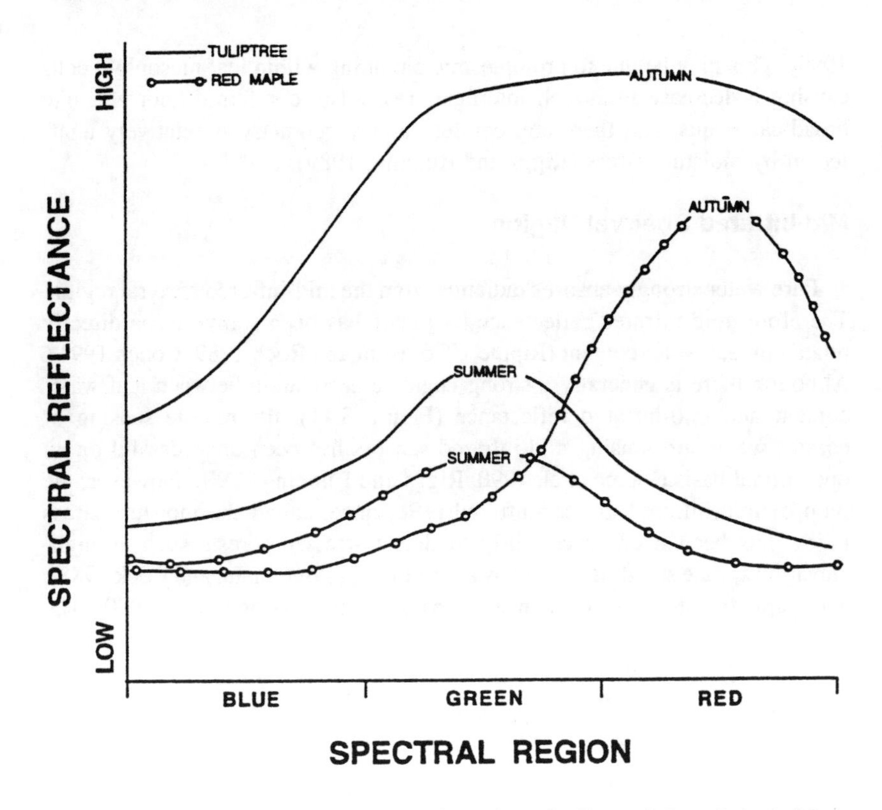

Figure 3.7. Visible spectral reflectance of maple and tuliptree foliage (adapted from Hoffer and Johannsen 1969).

Nearly any near-infrared radiation that is not reflected by a plant leaf will be transmitted. However, some of the transmitted near-infrared radiation will be reflected by a second, or a third layer of leaves. As forest canopies become denser, reflection of red light (due to increased chlorophyll absorbtion) decreases, while reflection of near-infrared radiation increases. Therefore, various red/near-infrared spectral indices can be used to estimate leaf area of closed-canopy conifer stands (Peterson et al. 1988). These red/near-infrared indices have also been used to map tropical deforestation in the Amazon basin (Woodwell et al. 1984), and to monitor seasonal leaf area dynamics of conifer stands (Curran et al. 1992).

Near-infrared leaf reflectance of broadleaf plants usually increases as leaf moisture-stress occurs (Thomas et al. 1966, Myers 1970). This increase in reflectance has been attributed to increased air/cell reflective surface area as internal cellular breakdown occurs (Ripple 1986). The increased near-infrared reflectance may also be due to increased exposure of highly reflective below-canopy substrate (such as dry leaf litter or soil) influencing reflectance measurements after plant canopy wilting (Curran and Milton 1983).

Near-infrared reflectance of conifer needles usually decreases with initiation of stress (Heller 1968, Rock et al. 1986, Westman and Price 1988, Murtha

1989). This may be due to protoplasmic clumping within leaf mesophyll cells causing a decrease in air/cell interface area. Also, conifers do not wilt like broadleaf plants, and therefore, conifer canopy geometry is relatively unaffected by moisture-stress (Riggs and Running 1989).

Mid-Infrared Spectral Region

Pure water strongly absorbs radiation from the mid-infrared spectral region. Therefore, mid-infrared reflectance by plants has been shown to be directly related to leaf water content (Ripple 1986, Hunt and Rock 1989, Cohen 1991). Although there is generally a strong negative correlation between leaf water content and mid-infrared reflectance (Figure 3.11), the remote sensing of conifer water stress using mid-infrared sensors has been unsuccessful on an operational basis (Pierce et al. 1990, Riggs and Running 1990). However, the ratio of mid-infrared to near-infrared reflectance, called the moisture stress index, has been used successfully to detect stressed forests such as high-elevation spruce stands damaged by air pollution (Vogelmann and Rock 1988), and maple forests damaged by pear thrips infestations (Vogelmann 1990). The

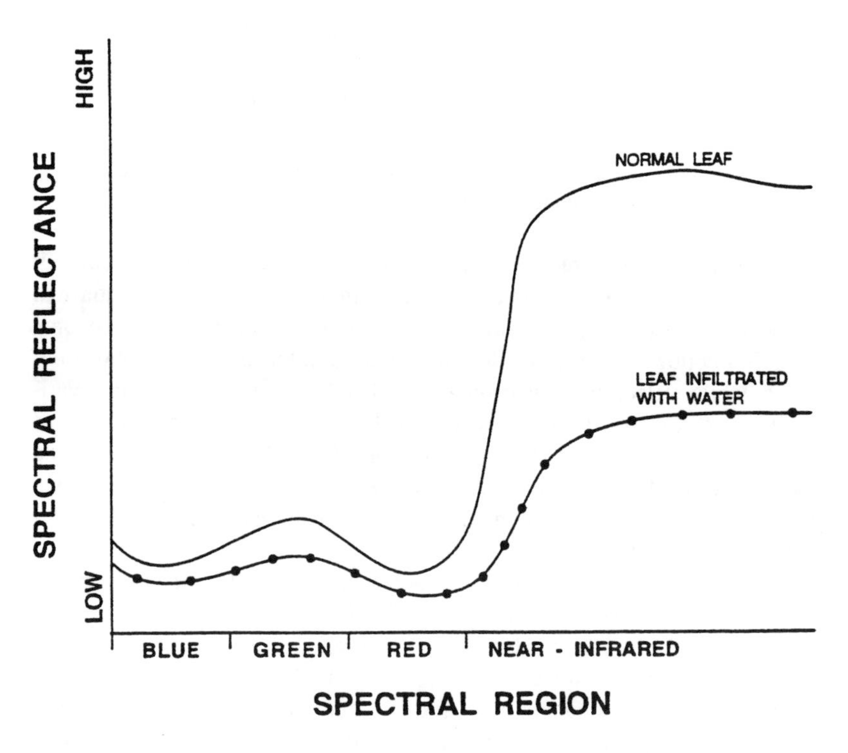

Figure 3.8. Decrease in near-infrared reflectance due to experimental decrease in cellular air gaps (adapted from Knipling 1970).

ratio of near-infrared to mid-infrared reflectance may also be sensitive to canopy shadowing conditions, and has been useful in remote sensing of old growth coniferous forests (Fiorella and Ripple 1993).

WATER SPECTRAL RELATIONSHIPS

If you were interested in the satellite remote sensing of aquatic vegetation, coral reefs, shoals, or other underwater features, you would need to use a spectral region that is high in transmittance through water. Transmittance is typically high in the blue-green spectral region and rapidly declines with longer wavelengths (Figure 3.12). Because of this spectral relationship, Landsat Thematic Mapper band 1 was designed for underwater mapping applications and therefore has a bandwidth of 0.45 to 0.52 μm.

Clear water reflects very little in most spectral regions. However, turbid water (water high in suspended solids) reflects significant amounts of radiation, especially in the red and near-infrared spectral regions. There is a shift in peak reflectance regions as water increases in turbidity (Figure 3.13), and

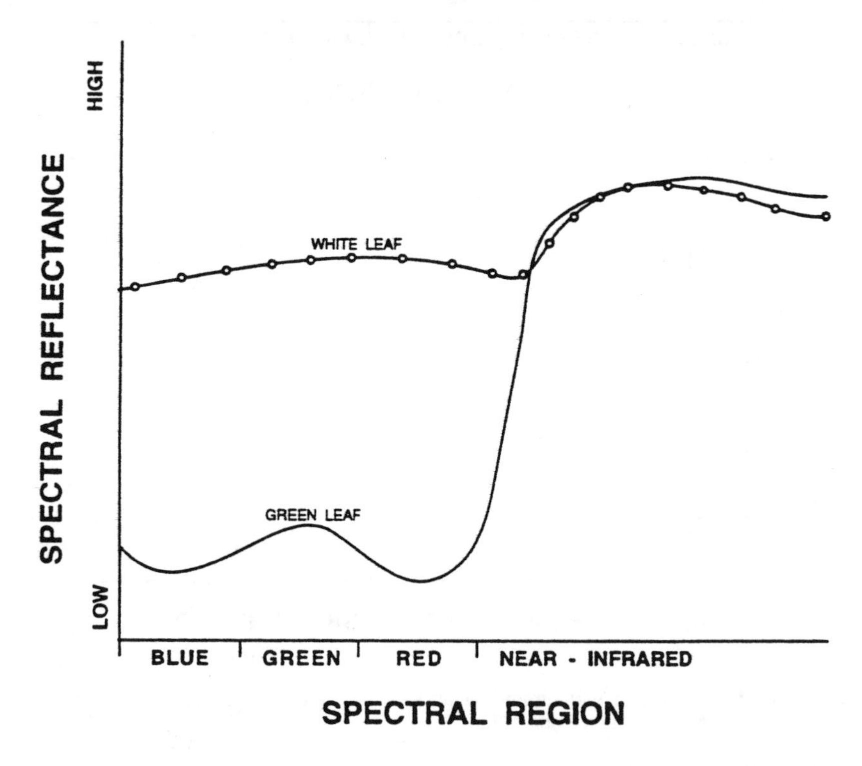

Figure 3.9. Spectral reflectance of white and green portions of a variegated geranium leaf (adapted from Billings and Morris 1951).

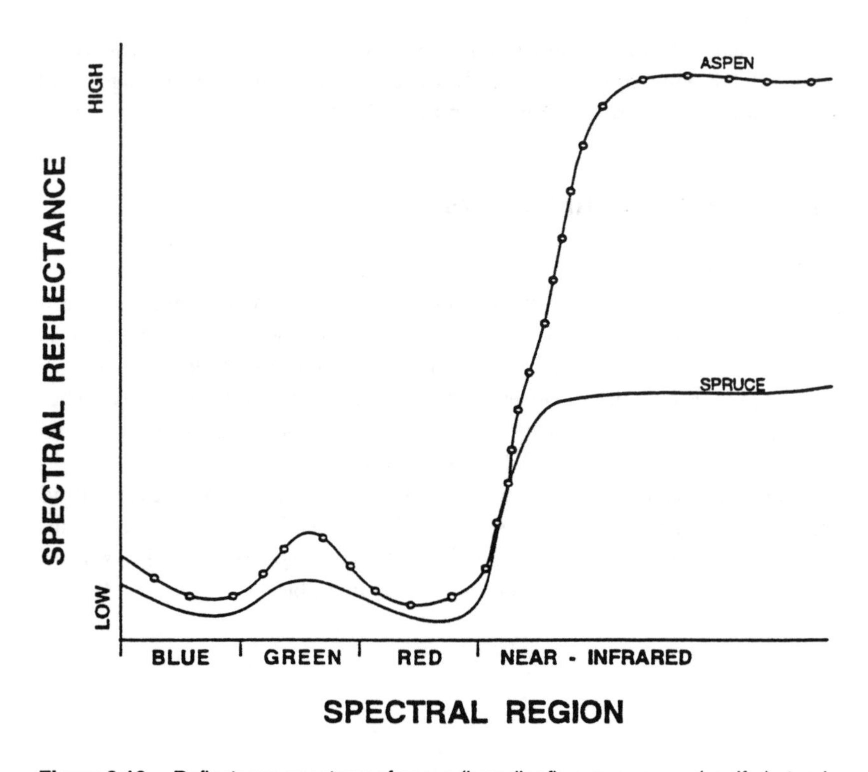

Figure 3.10. Reflectance spectrum of aspen (broadleaf) versus spruce (conifer) stands.

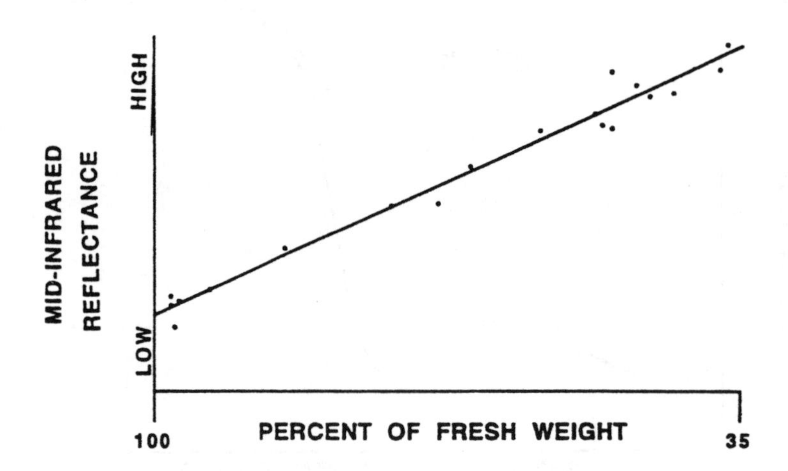

Figure 3.11. Change in mid-infrared reflectance as pine needles dry (adapted from Westman and Price 1988)

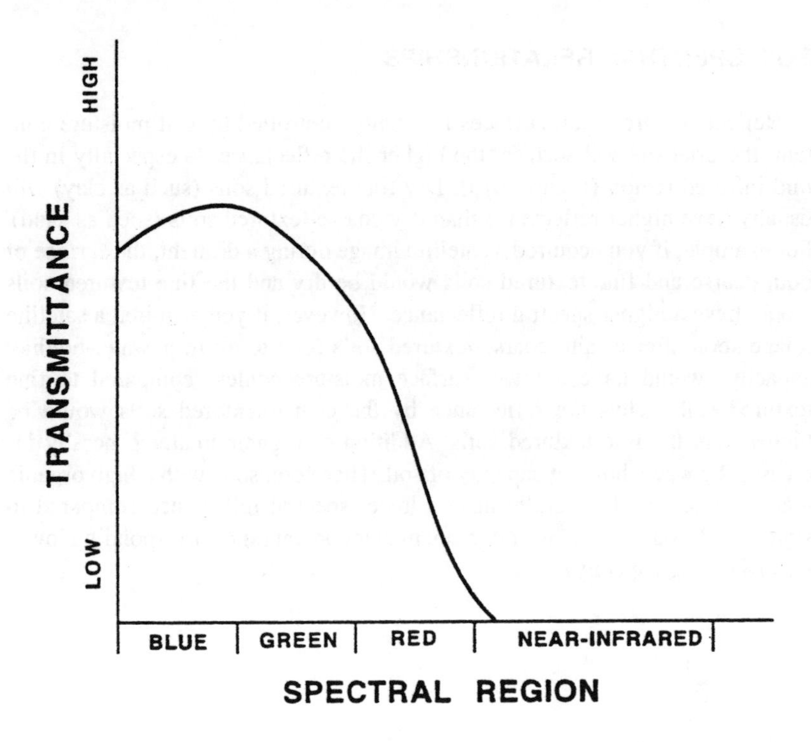

Figure 3.12. Spectral transmittance through clear water.

therefore, multi-band studies are usually made in monitoring water quality by satellite. Regression models are usually developed to predict suspended solid concentrations from spectral digital numbers. This approach has been successful in many studies (Moore 1980, Lathrop et al. 1986, Ritchie et al. 1990). However, the relationship between spectral reflectance and water quality may differ between water bodies. For example, Lathrop (1992) found that water quality models derived with Green Bay, Wisconsin spectral data could not be accurately applied to spectral data from Yellowstone and Jackson Lakes in Wyoming.

SNOW AND CLOUD SPECTRAL RELATIONSHIPS

It is usually difficult to distinguish snow versus clouds using satellite images from the visible spectral region. Clouds contain water droplets that are much larger than the visible wavelengths. Therefore, clouds scatter light nonselectively and appear white. Snow, on the other hand, has high reflectance in the visible and near-infrared regions and low reflectance in the mid-infrared spectral region. Therefore, Landsat Thematic Mapper band 5 (1.55 to 1.75 μm) is very useful for distinguishing between clouds and snow.

SOIL SPECTRAL RELATIONSHIPS

Reflectance from soil surfaces is usually controlled by soil moisture content: the drier the soil surface, the higher the reflectance — especially in the mid-infrared region (Figure 3.15). Dry fine-textured soils (such as clay) will usually have higher reflectance than dry coarse-textured soils (such as sand). For example, if you acquired a satellite image during a drought, the surface of both coarse and fine textured soils would be dry and the fine textured soils would have a higher spectral reflectance. However, if you acquired a satellite image soon after a rain, coarse-textured soils (due to a lower water-holding capacity) would have a lower surface moisture content compared to fine textured soils. Thus the reflectance by the coarse-textured soils would be higher than the fine-textured soils. Addition of organic matter generally increases the water-holding capacity of soil. Therefore, soils with a high organic matter content will generally have a lower spectral reflectance compared to similar soils that have a lower organic matter content (and corresponding lower water-holding capacity).

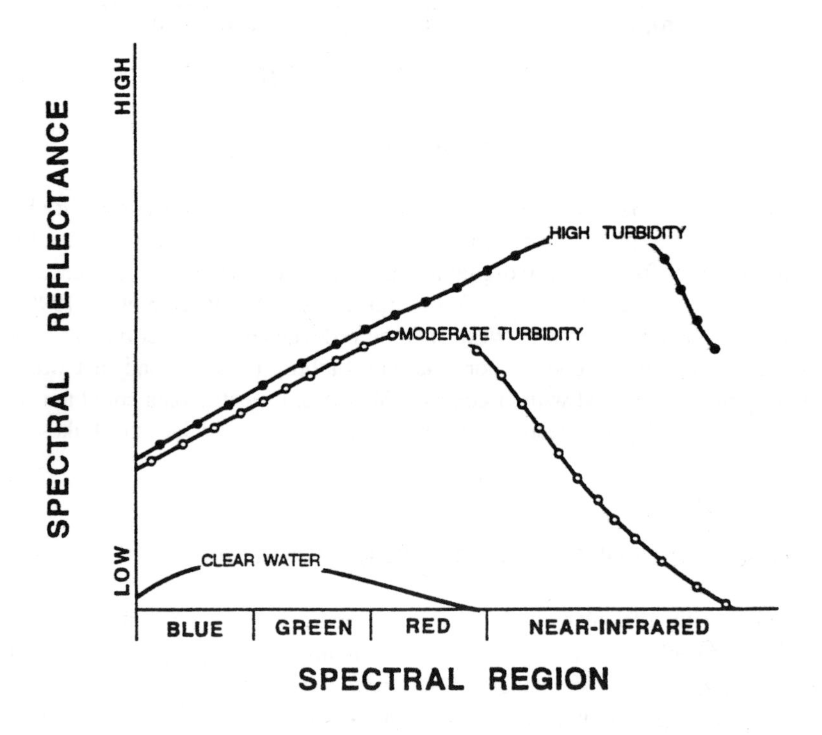

Figure 3.13. Spectral reflectance of clear and turbid water (adapted from Moore 1980).

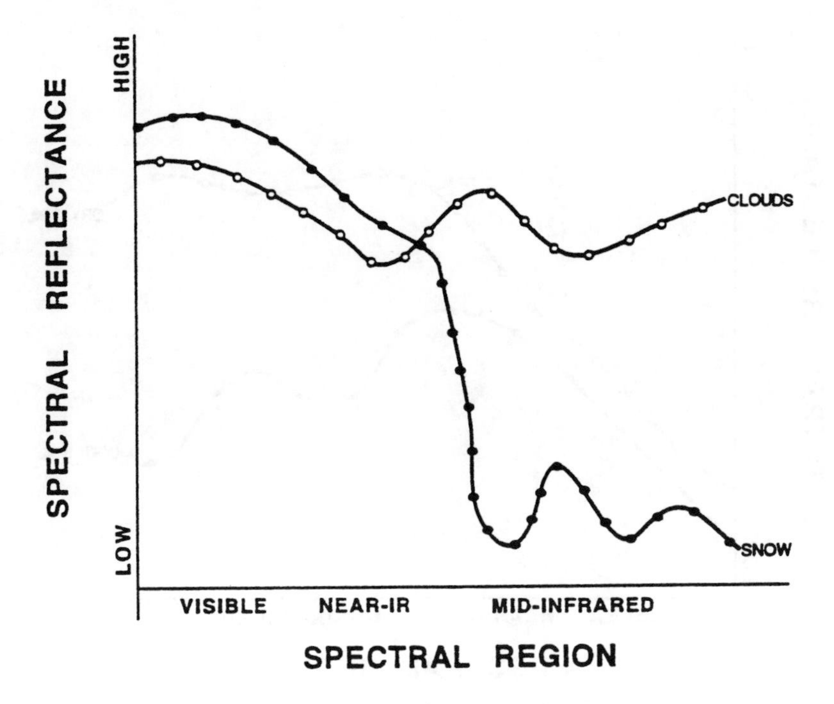

Figure 3.14. Spectral reflectance of clouds versus snow (adapted from Bowker et al. 1985).

THERMAL REMOTE SENSING

Any object with a temperature above absolute zero (–273°C) will emit (or radiate) radiation. On earth, the peak of this emitted radiation occurs in the thermal infrared region at wavelengths between 5 and 20 µm. Landsat Thematic Mapper detects thermal radiation from 10.4 to 12.4 µm (TM band 6) and has been used to estimate surface water temperatures within ±1°C (Gibbons et al. 1989). However, the radiant temperature that is measured by a sensor may not necessarily be the same temperature that is measured using a thermometer (termed kinetic temperature). An object will have the same radiant and kinetic temperatures only if it has an emissivity of 1; all energy received by the object is emitted back (termed a perfect blackbody radiator). Because most objects have an emissivity of less than 1, surface kinetic temperatures are usually recorded during satellite overpass and related to the relate remotely sensed thermal values. For example, if we were interested in surface water temperature mapping, we could use a thermometer to measure surface water temperatures along a transect during the satellite overpass. Later the satellite thermal values would be related to actual surface water temperature (Figure 3.16).

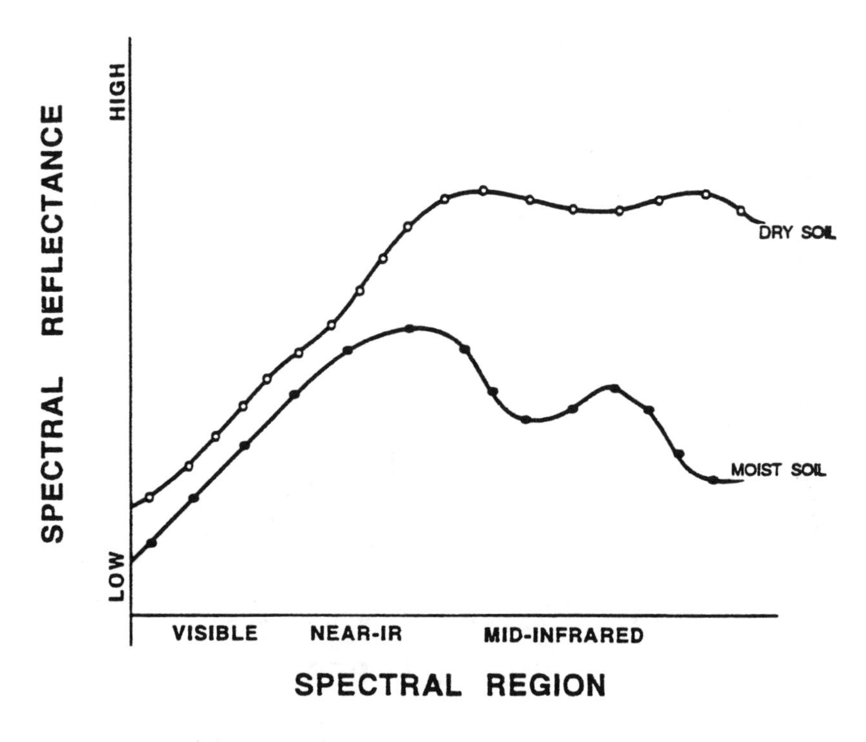

Figure 3.15. Spectral reflectance of dry versus moist soil.

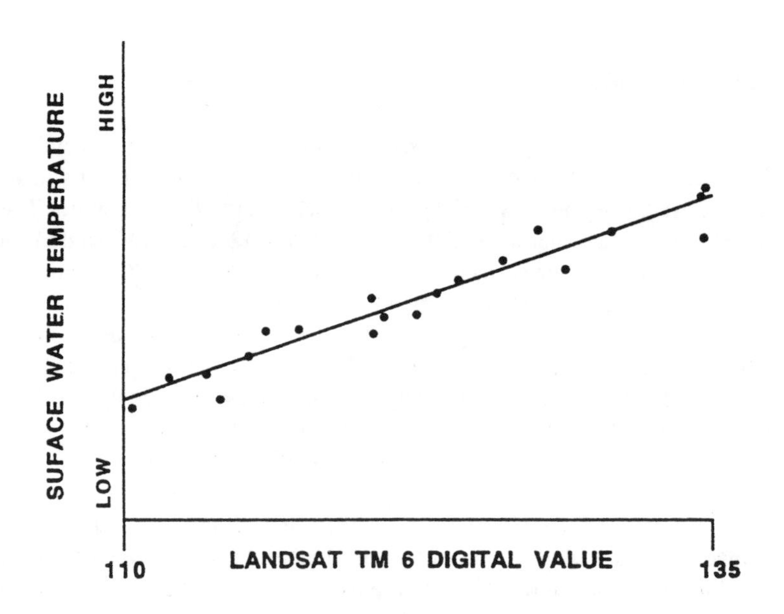

Figure 3.16. Relationship between surface water kinetic temperature and Landsat Thematic Mapper Band 6 Digital values (adapted from Lathrop and Lillesand 1986).

POTENTIAL PROBLEMS

A basic premise in satellite remote sensing is that cover types are spectrally different. By applying our understanding of spectral relationships, we can use satellite data for many natural resource applications. However, we will sometimes discover that known spectral relationships (often measured in the laboratory) are of limited value when used in field applications, because problems may occur that significantly affect the accuracy of our results. These potential problems include 1) mixed or polluted pixels, 2) spectral dominance, and 3) field conditions differing from laboratory conditions.

Mixed Pixels

A mixed or polluted pixel is a pixel that does not contain a homogeneous cover type. For example, imagine that we are mapping wetland cover types using one spectral band. We sample areas from the image that consist of homogeneous cover types and discover the following wetland types are spectrally different:

Wetland Cover Type	Digital Value Range
Sago Pondweed (SP)	30–39
Hardstem Bulrush (HB)	40–49
Cattail (CT)	50–59

Using the digital image in Figure 3.17, we might predict the wetland type for each pixel as shown.

We predicted the pixels that had digital values from 40 through 49 to be bulrush. However, some of these predictions may be incorrect because some of these pixels might be composed of a mixture of sago pondweed and cattail. For example, suppose that a pure stand of cattail has an average digital value of 55 and a pure stand of sago pondweed has an average value of 35. If we had

Digital Image:

35	36	38	40	41
37	38	39	42	43
38	39	40	43	54
39	41	45	53	55
41	43	46	54	57

Predicted Wetland Types:

SP	SP	SP	HB	HB
SP	SP	SP	HB	HB
SP	SP	HB	HB	CT
SP	HB	HB	CT	CT
HB	HB	HB	CT	CT

Figure 3.17. Predicted wetland classes assuming pure pixels.

Figure 3.18. Actual wetland classes with mixed pixels.

a pixel that was 50% cattail and 50% sago pondweed, the pixel might have a digital value of (0.5) (55) + (0.5) (35) = 45. This pixel might be incorrectly classified as bulrush rather than mixed cattail/pondweed. In fact, we might not have any bulrush in the wetland, and yet because of mixed pixels we might predict an extensive area of bulrush (Figure 3.18).

Spectral Dominance

In some applications, digital remote sensing may be difficult because digital values may be strongly influenced by factors that we are not interested in. For example, suppose we are interested in mapping sagebrush, four-winged salt-bush, and rabbitbrush range types. If these shrubs grow at a low density (less than 25% canopy coverage), it may be very difficult to spectrally differentiate these cover types because differences in spectral values may be primarily due to soil brightness differences and the vegetation type may have a minor influence on spectral values.

We might be interested in mapping chlorophyll concentrations from lake surface waters. However, this may be difficult in areas where runoff from agricultural fields is high because the concentration of suspended solids will then dominate spectral response (Ritchie et al. 1990).

Field Versus Laboratory Conditions

Spectral relationships discovered under controlled laboratory conditions might not be useful under some field conditions. For example, there have been many studies that have shown a very strong relationship between mid-infrared reflectance and plant moisture content (Westman and Price 1988, Hunt and Rock 1989, Cohen 1991), yet no studies have been able to detect moisture stress in conifers under operational conditions (Pierce et al. 1990, Riggs and Running 1990). This is because forest canopy variations, atmospheric attenuation, and other factors can mask the subtle spectral differences between moisture-stressed and control forest stands (Pierce et al. 1990).

Spectral relationships that are measured under laboratory conditions at the leaf level might be very different under field conditions at the canopy level. For example, Williams (1991) found that the magnitude of reflectance throughout the visible and near-infrared spectral regions decreased dramatically for conifer species as scene complexity increased from the individual needle level to the canopy level. This may be due to shadowing, absorbtion by twigs, branches and litter, and increased scattering of light. Understory cover can often strongly affect spectral reflectance values at the canopy level, especially in conifer stands, since the reflectance of understory grass, broadleaf shrubs, bedrock or soil is generally higher than the reflectance of conifers in most spectral regions. For example, Spanner et al. (1990) found a strong linear relationship between leaf area index and Landsat Thematic Mapper band 4 for conifer stands with greater than 89% canopy closure. The relationship for open stands with highly reflective broadleaf shrubs and/or soil backgrounds was much weaker.

Topography can also have a strong influence on spectral values. For example, Landsat passes over during midmorning when the sun is in the southeast (in the northern hemisphere). Imagine that we have an image with Douglas-Fir plantations on various slopes. Even if the plantations were previously thinned to the same basal area, the plantations on the southeast-facing slopes would have higher digital values than plantations on northwest-facing slopes. In the northeatern United States, the effect of topography is pronounced on images from fall, winter, and spring — times when deciduous trees are leafless and the angle of the sun is low. Pixels from northerly facing slopes on such images have low digital values and may be misclassified as wetlands, while pixels from southerly facing slopes have high digital values and may be misclassified as barren land (Civco 1989). Because of the strong influence of topography on spectral values, digital elevation data are often used with images from rugged terrain areas. In almost all cases, the use of digital elevation data has increased the accuracy of vegetation cover type mapping from satellite imagery (Cybula and Nyquist 1987, Frank 1988, Franklin and Moultan 1990).

Since satellites are orbiting hundreds of miles from the earth's surface, particulates and gases in the atmosphere can significantly affect spectral values by scattering or absorbing incoming and outgoing (reflected/emitted) radiation. For example, Spanner et al. (1990) found that for Landsat Thematic Mapper band 3, approximately 50% of total forest stand radiance was due to atmospheric path radiance (atmospheric scattering). Corrections for atmospheric effects will be discussed in the next chapter.

CHAPTER 3 PROBLEMS

1) Color film is sensitive to approximately the same wavelengths as the human eye (0.4 to 0.7 μm). List any satellite systems that have spectral bands that entirely cover this spectral region.

2) Color infrared aerial photography is often used in natural resource management for vegetation mapping. The color infrared film is exposed to wavelengths from 0.5 to 0.9 μm. List any satellite systems that have spectral bands that entirely cover this spectral region.

3) Plant leaves influence spectral patterns in many ways. Match the following Landsat Thematic Mapper bands with the appropriate influencing factor:

a) TM bands 1-3	____	reflection influenced mainly by moisture content.
b) TM bands 5, 7	____	reflection influenced mainly by plant pigments.
c) TM band 6	____	reflection influenced mainly by cell structure.
d) TM band 4	____	emittance influenced mainly by leaf temperature.

4) Suppose the following linear regression model was developed to predict surface water temperature in a large lake.

$$\text{Temperature} = -42.5 + 0.5(\text{Digitial Value})$$

Using the following thermal digital image, predict the surface water temperature for each grid cell.

133	130	129	129	128	127	125	123	120	118
131	129	128	128	127	128	126	122	119	116
129	129	128	128	127	125	122	119	118	115
127	126	125	125	125	125	123	119	117	113

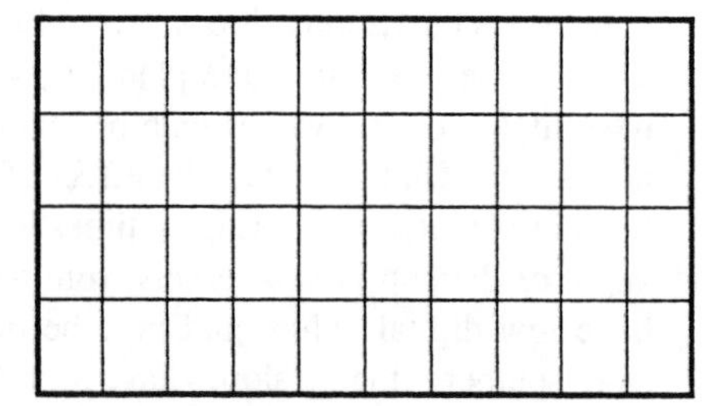

5) Read the following article: Pierce, L. L., Running, S. W. and G. A. Riggs. 1990. Remote Detection of canopy water stress in coniferous forests using the NS001 Thematic Mapper Simulator and the Thermal Infrared Multispectral Scanner. *Photogrammetric Engineering and Remote Sensing.* 56:579.586. Then answer the following questions:

As moisture stress occurs in a leaf, does the middle infrared reflectance increase or decrease? Why?

As moisture stress occurs in a leaf, does the thermal infrared emittance increase or decrease? Why?

As moisture stress occurs in broadleafs, near infrared reflectance also changes. Why does near infrared reflectance not change significantly in conifers as moisture stress occurs?

The density and species composition of overstory and understory plants can affect mid-infrared reflectance. What did the authors do to attempt to control for these potentially confounding factors?

Polluted pixels that occur along the border of a treatment or control stand can be a confounding factor. How did the authors deal with this potential problem?

Why would it be difficult to accurately map moisture stress in conifers using the Landsat Thematic Mapper mid-infrared bands?

6) There is a strong linear relationship between leaf area and the ratio of near-infrared/red reflectance. For example read:

   Running, S. W., Peterson, D. L., Spanner, M. A. and K. B. Teuber. 1986. Remote sensing of coniferous forest leaf area. *Ecology.* 67:273–276.

   Then read:

   Spanner, M. A., Pierce, L. L., Peterson, D. L. and S. W. Running. 1990. Remote sensing of temperate coniferous forest leaf area index: the influence of canopy closure, understory vegetation, and background reflectance. *International Journal of Remote Sensing.* 11:95–111.

   There was a strong relationship between near infrared reflectance and leaf area index of closed stands. Assume we developed a linear equation using this relationship. If we applied the model to the following stands, would we underestimate or overestimate leaf area index? Why?

   a) An old growth conifer stand with large trees and an understory of forest litter.
   b) An open canopy conifer stand with an understory of broadleaf shrubs or grass.
   c) An open canopy conifer stand with an understory of granite bedrock.
   d) A mixed conifer/broadleaf stand (75% conifer, 25% broadleaf).

7) Read the following:

   Ranson, J. Daughtry, C. S. T. and L. L. Biehl. 1986. Sun angle, view angle, and background effects on spectral response of simulated balsam fir canopies. *Photogrammetric Engineering and Remote Sensing.* 52(5):649–658.

   There was a very poor relationship between spectral indices and balsam fir biomass when the understory was grass and yet there was a very strong relationship when the understory was snow. Why?

   The authors conclude that "winter-time data, when snow masks the understory, would provide better estimates of overstory phytomass in coniferous forests than would summer time data." List at least one reason why exactly the opposite may be true with operational satellite remote sensing of balsam fir biomass.

8) Read the following:

   Lathrop, R. G. Jr. 1992. Landsat Thematic Mapper monitoring of turbid inland water quality. *Photogrammetric Engineering and Remote Sensing.* 58: 465–470.

Why is it difficult to estimate phytoplankton chlorophyll concentrates from satellite data of turbid inland lakes?

Why is it risky to extrapolate an existing remote sensing water quality model to regions that are different than the region where the model was developed?

LITERATURE CITED

Billings, W. D. and R. J. Morris. 1951. Reflection of visible and infrared radiation from leaves of different ecological groups. *American Journal of Botany.* 38:327–331.

Bowker, D. E., Davis, R. E., Myrick, D. L., Stacy, K. and W. T. Jones. 1985. *Spectral Reflectances of Natural Targets for Use in Remote Sensing Studies.* NASA Reference Publication 1139. 180 pp.

Cohen, W. B. 1991. Response of vegetation indices to changes in three measures of leaf water stress. *Photogrammetric Engineering and Remote Sensing.* 57:195–202.

Curran, P. J., Dungan, J. L., and H. L. Gholz. 1992. Seasonal LAI in slash pine estimated with Landsat TM. *Remote Sensing of Environment.* 39:3–13.

Fiorella, M. and W. J. Ripple. 1993. Determining the successional stage of temperate coniferous forests with Landsat satellite data. *Photogrammetric Engineering and Remote Sensing.* 59:239–246.

Gates, D. M. 1970. Physical and physiological properties of plants. In: *Remote Sensing with Special Reference to Agriculture and Forestry.* National Academy of Sciences, Washington, D. C. pp. 224–252.

Gausman, H. W., Escobar, D. E., and E. B. Knipling. 1977. Relation of Peperomia obtusifolia's anomalous leaf reflectance to its leaf anatomy. *Photogrammetric Engineering and Remote Sensing.* 43:1183–1185.

Heller, R. C. 1968. Previsual detection of ponderosa pine trees dying from bark beetle attack. *Proceedings of Fifth Symposium on Remote Sensing of the Environment,* Ann Arbor, MI. pp. 387–484.

Hoffer, R. M. and C. J. Johannsen. 1969. Ecological potentials in spectral signature analysis. In: *Remote Sensing in Ecology.* P. L. Johnson, ed., University of Georgia Press. pp. 1–16.

Hunt, E. R. Jr. and B. N. Rock. 1989. Detection of changes in leaf water content using near- and middle-infrared reflectances. *Remote Sensing of Environment.* 30:43–54.

Knipling, E. B. 1970. Physical and physiological basis for the reflectance of visible and near-infrared radiation from vegetation. *Remote Sensing of Environment.* 1:155–159.

Lathrop, R. G. Jr. 1992. Landsat Thematic Mapper monitoring of turbid inland water quality. *Photogrammetric Engineering and Remote Sensing.* 58:465–470.

Lathrop, R. G. Jr. and T. M. Lillesand. 1986. Utility of Thematic Mapper data to assess water quality in southern Green Bay and west-central Lake Michigan. *Photogrammetric Engineering and Remote Sensing.* 52:671–680.

Moore, G. K. 1980. Satellite remote sensing of water turbidity. *Hydrobiological Sciences.* 25:407–421.

Pierce, L. L., Running, S. W. and G. A. Riggs. 1990. Remote detection of canopy water stress in coniferous forests using the NS001 thematic mapper simulator and the thermal infrared multispectral scanner. *Photogrammetric Engineering and Remote Sensing.* 56:579–586.

Peterson, D. L., Westman, W. E., Stephensen, N. J., Ambrosia, V. G., Brass, J. A., and M. A. Spanner. 1988. Remote sensing of forest canopy and leaf biochemical contents. *Remote Sensing of Environment.* 24:85–108.

Author?? , Spanner, M. A., Running, S. W., and K. R. Teuber. 1987. Relationship of thematic mapper simulator data to leaf area index of temperate coniferous forests. *Remote Sensing of Environment.* 22:323–341.

Riggs, G. A. and S. W. Running. 1990. Detection of canopy water stress in conifers using the Airborne Imaging Spectrometer. *Remote Sensing of Environment.* 25:??.

Ripple, W. J. 1986. Spectral reflectance relationships to leaf water stress. *Photogrammetric Engineering and Remote Sensing.* 52:1669–1675.

Rock, B. N., Vogelmann, J. E., Williams, D. L., Vogelmann, A. F., and T. Hoshizaki. 1986. Remote detection of forest damage. *Bioscience.* 36:439–445.

Spanner, M. A., L. L Pierce, D. L. Peterson, and S. W. Running. 1990. Remote sensing of coniferous forest leaf area index: the influence of canopy closure, understory vegetation and background reflectance. *International Journal of Remote Sensing.* 11:95–111.

Tucker, C. J., Holben, B. N., and T. E. Goff. Intensive forest clearing in Rhondonia, Brazil, as detected by satellite remote sensing. *Remote Sensing of Environment.* 15:255–261.

Vogelmann, J. E. 1990. Comparison between two vegetation indices for measuring different types of forest damage in the north-eastern United States. *International Journal of Remote Sensing.* 12:2281–2297.

Vogelmann, J. E. and B. N. Rock. 1988. Assessing forest damage in high-elevation coniferous forests in Vermont and New Hampshire using thematic mapper data. *Remote Sensing of Environment.* 24:227–246.

Westman, W. E. and C. V. Price. 1988. Spectral changes in conifers subjected to air pollution and water stress: experimental studies. *IEEE Transactions on Geoscience and Remote Sensing.* 26:11–20.

Williams, D. L. 1991. A comparison of spectral reflectance properties at the needle, branch and canopy level for selected conifer species. *Remote Sensing of Environment.* 35:79–93.

ADDITIONAL READINGS

General Readings

Begni, G. 1982. Selection of the optimum spectral bands for the SPOT satellite. *Photogrammetric Engineering and Remote Sensing.* 48:1613–1620.

Coulson, K. L. and D. W. Reynolds. 1971. The spectral reflectance of natural surfaces. *Journal of Applied Meteorology.* 10:1285–1295.

Driscoll, R. S. 1993. Infrared demystified. *GIS World.* 6(2):32–34.

Hendricks, S. B. 1968. How light interacts with living matter. *Scientific American.* 219:175–186.

Lusch, D. P. 1989. Fundamental considerations for teaching the spectral reflectance characteristics of vegetation, soil and water. In: *Current Trends in Remote Sensing Education.* Nellis, M. D., Lougeay, R. and K. Lulla, eds. Geocarto International Centre, Hong Kong. 196 pp.

Malila, W. A. 1985. Comparison of the information contents of Landsat TM and MSS data. *Photogrammetric Engineering and Remote Sensing.* 51:1449–1457.

Weisskopf, V. F. 1968. How light interacts with matter. *Scientific American.* 219:60–71.

Vegetation Spectral Relationships

Ahern, F. J., Erdle, T., Maclean, D. A. and I. D. Kneppeck. 1991. A quantitative relationship between forest growth rates and Thematic Mapper reflectance measurements. *International Journal of Remote Sensing.* 12:387–400.

Curran, P. J. and E. J. Milton. 1983. The relationships between the cholorophyll concentration, LAI and reflectance of a simple vegetation canopy. *International Journal of Remote Sensing.* 4:247–255.

Danson, F. M. 1987. Preliminary evaluation of the relationships between SPOT-1 HRV data and forest stand parameters. *International Journal of Remote Sensing.* 8:1571–1575.

Dean, K. G., Kodama, Y. and G. Wendler. Comparison of leaf and canopy reflectance of subarctic forests. *Photogrammetric Engineering and Remote Sensing.* 52:809–811.

Fuller, S. P. and W. R. Rouse. 1979. Spectral reflectance changes accompanying a post-fire recovery sequence in a subarctic spruce lichen woodland. *Remote Sensing of Environment.* 8:11–23.

Gates, D., Keegan, H. J., Schelter, J. C., and V. R. Weidner. 1965. Spectral properties of plants. *Applied Optics.* 4:11–20.

Gausman, H. W. 1974. Leaf reflectance of near-infrared. *Photogrammetric Engineering and Remote Sensing.* 60(2):183–191.

Graetz, R. D. and M. R. Gentle. 1982. The relationships between reflectance in the Landsat wavebands and the composition of an Australian semi-arid shrub rangeland. *Photogrammetric Engineering and Remote Sensing.* 48:1721–1730.

Guyot, G., Guyon, D. and J. Riom. 1989. Factors affecting the spectral response of forest canopies: a review. *Geocarto International.* 3–18.

Howard, J. A. 1991. *Remote Sensing of Forest Resources: Theory and Applications.* Chapman & Hall, London. 420 pp.

Jackson, R. D. 1986. Remote sensing of biotic and abiotic plant stress. *Annual Review of Phytopathology.* 24:265–287.

Tucker, C. J. 1980. Remote sensing of leaf water content in the near infrared. *Remote Sensing of Environment.* 10:23–32.

Tucker, C. J., Holben, B. N., and T. E. Goff. 1984. Intensive forest clearing in Rhondonia, Brazil, as detected by satellite remote sensing. *Remote Sensing of Environment.* 15:255–261.

Wooley, J. T. 1971. Reflectance and transmittance of light by leaves. *Plant Physiology.* 47:656–662.

Water Spectral Relationships

Bartolucci, L. A., Robinson, B. F. and L. F. Silva. 1977. Field measurements of the spectral response of natural waters. *Photogrammetric Engineering and Remote Sensing.* 43:595–598.

Bhargava, D. S. and D. W. Mariam. 1990. Spectral reflectance relationships to turbidity generated by different clay materials. *Photogrammetric Engineering and Remote Sensing.* 56:225–229.

Curran, P. J., and E. M. M. Novo. 1988. The relationship between suspended sediment concentration and remotely sensed spectral radiance: a review. *Journal of Coastal Research.* 4:351–368.

Witzig, A. S. and C. A. Whitehurst. 1981. Current use and technology of Landsat MSS data for lake trophic classification. *Water Resources Bulletin.* 17:962–970.

Snow and Cloud Spectral Relationships

Derrien, M., Farki, B., Harang, L., LeGleau, H., Noyalet, A., Pochic, D., and A. Sairouni. 1993. Automatic cloud detection applied to NOAA-11/AVHRR imagery. *Remote Sensing of Environment.* 46:246–267.

Dozier, J., Schneider, S. R., and D. F. McGinnis, Jr. 1981. Effect of grain size and snowpack water equivalence on visible and near-infrared satellite observations of snow. *Water Resources Research.* 17:1213–1221.

Dozier, J., 1989. Spectral signature of alpine snow cover from the Landsat Thematic Mapper. *Remote Sensing of Environment.* 28:9–22.

Eck, T. F. and V. L. Kalb. 1991. Cloud-screening for Africa using a geographically and seasonally variable infrared threshold. *International Journal of Remote Sensing.* 12:1205–1221.

Hall, D. K., Foster, J. L., and A. T. C. Chang. 1992. Reflectance of snow as measured in situ and from space in sub-arctic areas in Canada and Alaska. *IEEE Transactions on Geoscience and Remote Sensing.* 30:634–637.

Soil Spectral Relationships

Ben-Dor, E. and A. Banin. 1994. Visible and near-infrared (0.4–1.1 μm) analysis of arid and semiarid soils. *Remote Sensing of Environment.* 48:261–274.

Bowers, S. S. and R. J. Hanks. 1965. Reflection of radiant energy from soils. *Soil Science.* 100:130–138.

Condit, H. R. 1980. The spectral reflectance of American soils. *Photogrammetric Engineering and Remote Sensing.* 36:955–966.

Csillag, F., Pasztor, L., and L. L. Biehl. 1993. Spectral band selection for the characterization of salinity status of soils. *Remote Sensing of Environment.* 43:231–242.

Huete, A. R. 1988. A soil-adjusted vegetation index (SAVI). *Remote Sensing of Environment.* 25:295–309.

Kimes, D. S., Irons, J. R., Levine, E. R. and N. A. Horning. 1993. Learning class descriptions from data base of spectral reflectance of soil samples. *Remote Sensing of Environment.* 43:161–169.

Price, J. C. 1990. On the information content of soil reflectance spectra. *Remote Sensing of Environment.* 33:113–121.

Thermal Remote Sensing

Baur, B. and A. Baur. 1993. Climatic warming due to thermal radiation from an urban area as possible cause for the local extinction of a land snail. *Journal of Applied Ecology.* 30:333–340.

Cihlar, J. 1980. Soil, Water and plant canopy effects on remotely measured surface temperatures. *International Journal of Remote Sensing.* 1:167–173.

Gibbons, D. E., Wukelic, G. E., Leighton, J. P. and M. J. Doyle. 1989. Application of Landsat Thematic Mapper data for coastal thermal plume analysis at Diablo Canyon. *Photogrammetric Engineering and Remote Sensing.* 55:903–909.

LeDrew, E. F. and S. E. Franklin. 1985. The use of thermal infrared imagery in surface current analysis of a small lake. *Photogrammetric Engineering and Remote Sensing.* 51:565–573.

Moran, M. W., Jackson, R. D., Raymond, L. H., Gay, L. W. and P. N. Slater. 1989. Mapping surface energy balance components by combining Landsat Thematic Mapper and ground-based meteorological data. *Remote Sensing of Environment.* 30:77–87.

Robinson, J. M. Fire from space: global fire evaluation using infrared remote sensing. *International Journal of Remote Sensing.* 12:3–24.

Wukelic, G. E., Gibbons, D. E., Martucci, L. M. and H. P. Foote. 1989. Radiometric calibration of Landsat Thematic Mapper Thermal Band. *Remote Sensing of Environment.* 28:339–347.

Yates, H. W. Thermal infrared in remote sensing: a historical review. *Technical Papers 1989 ASPRS / ASCM Annual Convention.* pp. 212–216.

Topographic Corrections

Cavayas, F. 1987. Modelling and correction of topographic effect using multi-temporal satellite images. *Canadian Journal of Remote Sensing.* 13:49–67.

Civco, D. L. 1989. Topographic normalization of Landsat TM imagery. *Photogrammetric Engineering and Remote Sensing.* 55:1303–1309.

Colby, J. D. 1991. Topographic normalization in rugged terrain. *Photogrammetric Engineering and Remote Sensing.* 57:531–537.

Cybula, W. G. and M. O. Nyquist. 1987. Use of topographic and climatological models in a geographic database to improve Landsat MSS classification for Olympic National Park. *Photogrammetric Engineering and Remote Sensing.* 53:59–65.

Egbert, D. D. and F. T. Ulaby. 1972. Effect of angles on reflectivity. *Photogrammetric Engineering and Remote Sensing.* 29:556–564.

Frank, T. D. 1988. Mapping dominant vegetation communities in the Colorado Rocky Mountain Front Range with Landsat Thematic Mapper and digital terrain data. *Photogrammetric Engineering and Remote Sensing.* 54:1727–1734.

Franklin, J., Logan, T., Woodcock, C. and A. Strahler. 1986. Coniferous forest classification and inventory using Landsat and digital terrain data. *IEEE Transactions on Geosciences and Remote Sensing.* GE-24:139–149.

Guindon, B., Goodenough, D. G. and P. M. Teillet. 1982. The role of digital terrain models in the remote sensing of forests. *Canadian Journal of Remote Sensing.* 8:4–16.

Hall-Konyves, K. 1987. The topographic effect on Landsat data in gently unulating terrain in southern Sweden. *International Journal of Remote Sensing.* 8:157–168.

Holben, B. N. and C. O. Justice. 1980. Topographic effect on spectral response from nadir-pointing sensors. *Photogrammetric Engineering and Remote Sensing.* 46:1191–1200.

Holben, B. N. and C. O. Justice. 1981. An examination of spectral band ratioing to reduce the topographic effect on remotely sensed data. *International Journal of Remote Sensing.* 2:115–133.

Huchinson, C. F. 1982. Techniques for combining landsat and ancillary data for digital classsification improvement. *Photogrammetric Engineering and Remote Sensing.* 48:122–130.

Justice, C. S., Wharton, S. W., and B. N. Holben. 1981. Application of digital terrain data to quantify and reduce the topographic effect on Landsat data. *International Journal of Remote Sensing.* 2:213–230.

Kimes, D. S., Smith, J. A., and K. J. Ranson. 1980. Vegetation reflectance measurements as a f function of solar zenith angle. *Photogrammetric Engineering and Remote Sensing.* 46:15631573.

Leprieur, C. E. and J. M. Durand. 1988. Influence of topography on forest reflectance using Landsat TM and digital terrain data. *Photogrammetric Engineering and Remote Sensing.* 54:491–496.

Proy, C., Tanre, D. and P. V. Deschamps. 1989. Evaluation of topographic effects on remotely sensed data. *Remote Sensing of Environment.* 30:21–32.

Shasby, M. and D. Carneggie. 1986. Vegetation and terrain mapping in Alaska using Landsat MSS and digital terrain data. *Photogrammetric Engineering and Remote Sensing.* 52:779–786.

Teillet, P. M., Guindon, B. and D. Goodenough. 1982. On the slope-aspect correction of multispectral scanner data. *Canadian Journal of Remote Sensing.* 8:84–106.

CHAPTER 4

Radiometric Corrections

INTRODUCTION

Radiometric corrections are sometimes used to adjust digital numbers as a preprocessing step before an image is classified. In this chapter, you will learn several simple methods that can be applied for radiometric corrections of errors due to: 1) faulty sensor operation, and 2) atmospheric scattering of radiation. There are many more complex methods for radiometric corrections that are beyond the scope of this primer. Some of these methods are cited in the *Literature Cited* and *Additional Readings* section of this chapter.

DETECTOR ERRORS

In most digital remote sensors, each detector senses radiance along one line. For example, Landsat Thematic Mapper has 16 detectors for each reflective band and 4 detectors for the thermal (emissive) band. The reflection detectors produce a 16-line portion of an image with each scan (Figure 4.1).

Line Dropout

Sometimes a digital image contains bad scan lines that are black (or contain pixel values of zero). This condition is often called *line dropout.* Line dropout can be due to detector errors, errors in transmitting the image data to a ground

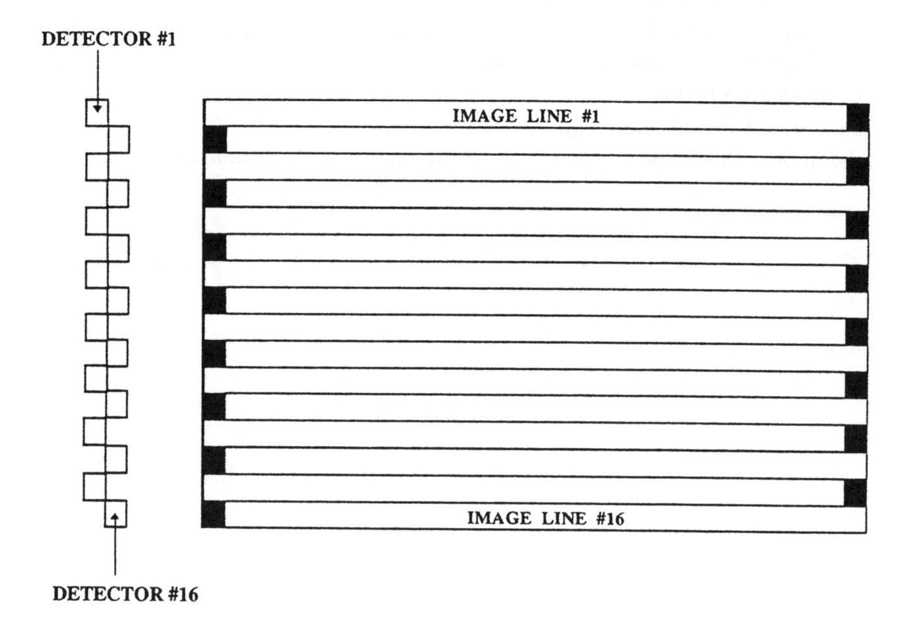

Figure 4.1. Landsat Thematic Mapper detector system.

78	79	74	75
77	78	71	73
0	0	0	0
74	73	69	69

78	79	74	75
77	78	71	73
76	76	70	71
74	73	69	69

Figure 4.2. Line dropout due to malfunctioning detector #3.

receiving station, or errors in processing the image data and writing it to a computer tape. Line dropout can be corrected by simply computing the mean of the pixel above and below each bad pixel (Figure 4.2). This approach is *only* cosmetic, but it improves the visual appearance of the image. However, you should be aware that the data in the corrected scan line are manufactured from surrounding scan lines, and are not data that were recorded by the satellite sensor.

Another method of replacing bad scan lines is to use highly correlated values from other spectral bands to predict the pixel values of the bad scan lines. Fusco and Trevese (1985) found that this method was more accurate than the simpler pixel averaging adjustment.

Destriping

A systematic banding or striping pattern can sometimes be found on images produced by scanning mirrors such as Landsat MSS or Thematic Mapper. This striped pattern is usually most evident in dark homogeneous areas of an image such as water bodies or dense forest areas. The problem occurs when an individual detector's radiometric response drifts from its initial (prelaunch calibration) setting. For example, each Landsat MSS band has 6 detectors. If one of the detectors (let's say detector #3, for example) becomes less sensitive to incoming radiance, the digital numbers along that can line will be consistently less than the digital numbers from the other five detectors. The image would then appear to have darker stripes at lines 3, 9, 15, 21, and so on.

Destriping procedures generally produce a histogram from scan lines corresponding to each detector. For example, we could generate a histogram for detector #1 by using the digital values from line 1, 7, 13, etc. We could do the same to generate a histogram for detector #2 using digital values from line 2, 8,14, etc. After we have generated a histogram corresponding to each of the 6 detectors, the histograms could be used to determine unusual detector responses. Correction of digital values from the unusual detector response could then be accomplished by determining an appropriate value to shift the histogram to match the average histogram from the normal detectors (Ahern et al. 1987).

CORRECTION FOR ATMOSPHERIC SCATTERING

Since satellite data are recorded hundreds of miles from the earth's surface, particulate (aerosols) and gases in the atmosphere can scatter, absorb, and refract radiation as it travels the path from the earth's surface to the sensor. The most dominant atmospheric effect is usually scattering of radiation by particulate, especially in the visible wavelengths (Chavez 1988). Therefore, satellite images are sometimes preprocessed to adjust for atmospheric effects.

Typical Applications

There have been hundreds of applications in which satellite remote sensing of natural resources was accomplished without correction for atmospheric effects. However, there are at least four types of applications where correction for atmospheric effects are extremely important: 1) aquatic applications, 2) ratio adjustments, 3) multi-temporal applications, and 4) multi-sensor applications.

Aquatic Applications

In aquatic applications, most of the solar radiation is transmitted or absorbed by water bodies. A very small fraction of solar energy is backscattered from the surface of clear, deep water. Therefore, much of the radiance detected by a satellite sensor is due to atmospheric scattering of radiation. More than 95% of the radiance received by a satellite sensor over clear, deep water may be due to atmospheric scattering (Moore 1980). Without correcting for atmospheric scattering of radiation, it may not be possible to detect surface water conditions such as chlorophyll concentration. The problem is analogous to attempting to take clear photographs on a foggy or smoky day. In fact, corrections for atmospheric scattering are often referred to as *dehazing corrections.*

Ratio Adjustments

Atmospheric scattering of solar radiation usually varies inversely with wavelength. For example, in the visible spectral region, blue wavelengths scatter much more than longer wavelengths such as green or red light. Because the scattering of short wavelength blue light dominates, the sky appears blue.

Imagine we wanted to use the ratio of red to near-infrared radiance values to estimate leaf area index. The red digital numbers would be inflated due to atmospheric scattering. Therefore, the leaf area ratio would be biased. Spanner et al. (1990) found that the path radiance (radiance mostly due to atmospheric scattering) contributed to about 50% of the radiance in Thematic Mapper band 3 (red) and 20% in Thematic Mapper band 4 (near-infrared). They concluded that correction of TM data for atmospheric effects is necessary for accurate regional leaf area index estimation.

Multi-Image Applications

Imagine that we are planning to subtract Landsat MSS images from 1972 and 1990 to determine areas that have changed within a wilderness area. In order for this change detection method to work, pixels containing the same cover type in 1972 and 1990 should have nearly the same digital values. However, the atmospheric "haze" from the 1972 and 1990 image probably is not the same. Therefore it is desirable to apply corrections for atmospheric scattering to both images before a change detection method is applied.

Multi-Sensor Applications

Some studies combine imagery from different sensors. For example, we might want to use both SPOT HRV multispectral and Landsat Thematic Mapper imagery to maximize the number of available cloud-free images for a water quality study. In order to make valid comparisons between the two images, we must correct for differences between the images due to atmospheric scattering.

Haze Remove Strategies

Several different atmospheric scattering or "haze" removal techniques have been developed. In general, these techniques can be grouped into three categories: 1) methods using complex atmospheric transmission models requiring field or atmospheric information taken during the time of satellite overflight (Forster 1984, Spanner et al. 1990, Tanre et al. 1990), 2) methods that use images from two or more dates for atmospheric corrections (Caselles and Garcia 1989, Hall et al. 1991), and 3) methods that estimate the effect of atmospheric scattering using the image data (Switzer et al. 1981, Crist 1984, Chavez 1988, Rice and Odenweller 1990, Lavreau 1991).

Histogram Adjustment Technique

The simplest approach for adjusting image data due to the additive bias from atmospheric scattering is often called the *histogram adjustment* technique (Chavez 1988). This approach is commonly used in current research (e.g., Nel et al. 1994, Nemani et al. 1993). The basic assumption is that there is a high probability that at least a few pixels in a satellite scene should have digital values of zero. This assumption is made because of the great number of pixels in any one scene (for example a Landsat MSS scene contains more than 10,000,000 pixels, and a Landsat TM scene contains more than 45,000,000 pixels). Therefore, there is usually some shadow areas (due to canyon shadow or cloud shadow) or deep, clear water in the image where the digital values should be zero. However, because of the additive effect of atmospheric scat-

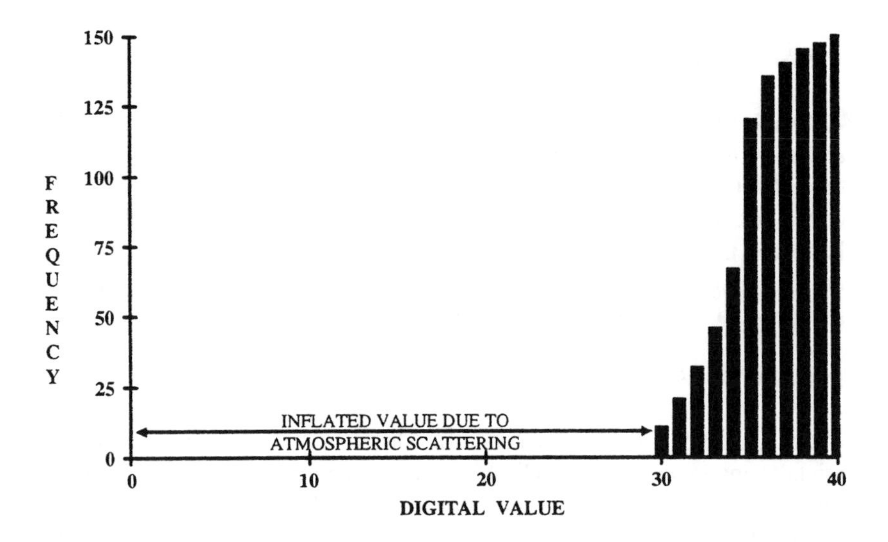

Figure 4.3. Lower end of histogram from a Landsat TM band 1 scene.

tering (haze), the digital values in these areas may indeed be greater than zero. For example, if the digital numbers in dark, shaded areas of a scene were 30, we would assume that this is due to atmospheric scattering and all digital numbers in this particular band should be adjusted by –30. The method is called the *histogram adjustment* technique; it shifts the histogram of digital numbers by a constant which is assumed to be the additive bias due to atmospheric scattering (Figure 4.3).

You should always examine the location of "dark pixels" before assuming that they contain true "dark digital values." This is because, in some cases, the detector may be saturated by high reflectance from a bright area. The next few pixels along the scan line may then be negatively biased (lower digital numbers). Some common examples include: 1) bright cloud pixels followed by a cloud shadow pixel, 2) bright sandy beach pixels followed by a water pixel, and 3) bright southeast facing slope pixels followed by a shadow pixel on a northwest-facing slope. An analogy of this negative bias is the condition we perceive as "snow-blindness." If you enter a room after skiing or hiking in bright sunny conditions, your eyes as remote sensors are temporarily negatively biased and the room looks darker than normal.

The histogram adjustment technique is simple but has two potential problems. If histograms from a small study area (rather than a full scene) are used, a minimum digital value from the full scene may not appear in the histogram. Therefore the minimum digital value selected would overcorrect for atmospheric haze. Additionally, The method may cause some digital values to be overcorrected in some bands and under-corrected in other bands. If this occurs, the spectral relationships between bands may be distorted. Chavez (1988) has developed an improved dark-object subtraction technique that is beyond the scope of this primer.

CHAPTER 4 PROBLEMS

1) Find two remote sensing articles that applied some sort of correction for atmospheric scattering and two articles that did not use any correction. You might find the articles in the following periodicals: Remote Sensing of Environment, Photogrammetric Engineering and Remote Sensing, International Journal of Remote Sensing, ISPRS Journal of Photogrammetry and Remote Sensing, GeoCarto International, GIS World, ASPRS-ASCM Convention Proceedings, etc.

   List the complete citation from each paper including all authors, date, periodical, volume and pages. Discuss why the authors decided to either apply or not apply atmospheric correction to the images.

2) Read the paper by Robinove (1982) on computing physical values from digital numbers. Then list one periodical article that actually used physical values rather than digital numbers.

   List the complete citation from each paper including all authors, date, periodical, volume and pages. Discuss why your selected study used physical values instead of digital numbers.

LITERATURE CITED

Ahern, F. J., Brown, R. J., Cihlar, J., Gauther, R., Murphy, J., Neville, R. A., and P. M. Teillet. 1987. Radiometric correction of visible and infrared remote sensing data at the Canada Centre for Remote Sensing. *International Journal of Remote Sensing*. 8:1349–1376.

Caselles, V. and M. J. Lopez Garcia. 1989. An alternative simple approach to estimate atmospheric correction in multitemporal studies. *International Journal of Remote Sensing*. 10:1127–1134.

Chavez, P. S. Jr. 1988. An improved dark-object subtraction technique for atmospheric scattering correction of multispectral data. *Remote Sensing of Environment*. 24:459–479.

Crist, E. P. 1984. A spectral haze diagnostic feature for normalizing Landsat thematic mapper data. *Proceedings of the 18th International Symposium on Remote Sensing of the Environment.* Vol. 18. pp 735–745.

Forster, B. C. 1984. Derivation of atmospheric correction procedures for Landsat MSS with particular reference to urban data. *International Journal of Remote Sensing*. 5:799–817.

Fusco, L. and Trevese, D. 1985. On the reconstruction of lost data in images of more than one band. *International Journal of Remote Sensing*. 6:1535–1544.

Hall, F. G., Strebel, D. E., Nickeson, J. E. and S. J. Goetz. 1991. Radiometric rectification: Towards a common radiometric response among mutidate, multisensor images. *Remote Sensing of Environment.* 35:11–27.

Lavreau, J. 1991. De-hazing Landsat Thematic Mapper images. *Photogrammetric Engineering and Remote Sensing*. 57:1297–1302.

Moore, G. K. 1980. Satellite remote sensing of water turbidity. *Hydrobiological Sciences.* 25:407–421.

Nel, E. M., Wessman, C. A., and T. T. Veblen. 1994. Digital and visual analysis of Thematic Mapper imagery for differentiating old growth from younger spruce-fir stands. *Remote Sensing of Environment.* 48:291–301.

Nemani, R., Pierce, L. Running, S. and S. Goward. 1993. Developing satellite-derived estimates of surface moisture status. *Journal of Applied Meteorology.* 32:548–557.

Robinove, C. J. 1982. Computation with physical values from Landsat digital data. *Photogrammetric Engineering and Remote Sensing.* 48:781–784.

Spanner, M. A., Pierce, L. L., Peterson, D. L., and S. W. Running. 1990. Remote sensing of temperate coniferous forest leaf area index: The influence of canopy closure, understory vegetation and background reflectance. *International Journal of Remote Sensing.* 11:95–11.

Switzer, P., Kowalick, W. S., and R. J. P. Lyon. 1981. Estimation of atmospheric path-radiance by the covariance matrix method. *Photogrammetric Engineering and Remote Sensing.* 47:1469–1476.

Tanre, D., Deroo, C., Duhaut, P. Herman, M., Morcrette, J. J., Perbos, J., and P. Y. Deschamps. 1990. Description of a computer code to simulate the satellite signal in the solar spectrum: the 5S code. *International Journal of Remote Sensing.* 2:659–668.

ADDITIONAL READINGS

Ahern, F. J., Goodenough, S. J., Rao, V. and G. Rochon. 1977. Use of clear lakes as standard reflectors for atmospheric measurements. *Proceedings of the 11th International Symposium on Remote Sensing of the Environment.* Vol. 1. pp. 731–755.

Chavez, P. S. Jr. 1989. Radiometric calibration of Landsat thematic mapper multispectral images. *Photogrammetric Engineering and Remote Sensing.* 55:1285–1294.

Markham, B. L. and J. L. Barker. 1985. Spectral characterization of the Landsat Thematic Mapper sensors. *International Journal of Remote Sensing.* 6:697–716.

Price, J. C. 1987. Calibration of satellite radiometers and the comparison of vegetation indices. *Remote Sensing of Environment.* 21:15–27.

Rice, D. P. and J. B. Odenweller. 1990. External effects correction of Landsat Thematic Mapper data. *ISPRS Journal of Photogrammetry and Remote Sensing.* 44:355–368.

Verdin, J. P. 1985. Monitoring water quality conditions in a large western reservoir with Landsat imagery. *Photogrammetric Engineering and Remote Sensing.* 51:343–353.

CHAPTER 5

Geometric Corrections

INTRODUCTION

Digital images of earth's surface are distorted. *Rectification* (also called rubber sheeting) is the process of removing distortion from imagery by warping the image to fit a map projection. Each pixel is usually assigned a map coordinate during rectification. Then the rectified image can be used as a planimetric map, for on-screen digitizing with a GIS, or for image classification.

MAP PROJECTIONS

A map projection is a mathematical system of projecting a spheroid-shaped planet (such as the earth) onto a flat plane. It is *impossible* to project the spheroid earth onto a flat piece of paper and preserve accurate map properties of shape, area, distance and direction. Therefore there are hundreds of map projections that compromise these map properties. For example, you may have noticed that countries such as Greenland and Canada appear very large on some world maps and appear smaller on other world maps. Each map projection system has certain advantages and limitations. Several common projection systems used in digital remote sensing of natural resources include the Mercator Projection, the Transverse Mercator Projection, and the Space Oblique Mercator Projection.

Mercator Projection

This projection was introduced in the 1500s by a Flemish cartographer named Mercator. The Mercator projection is basically a cylinder tangent to the equator (Figure 5.1). This projection is analogous to fitting a tin can over an orange and having the tin can touch exactly at the "equator" of the orange. Mercator devised this map projection specifically for navigation; all lines of constant compass bearing appear as straight lines on maps using this projection, a pretty nice feature if you were trying to cross the ocean in a ship during the 1500s.

Transverse Mercator Projection

Because the projection "touches" along the equator, the Mercator projection has an exact scale at the equator and larger scale toward the poles. Therefore, it is more accurate in the east-west direction than the north-south direction. The

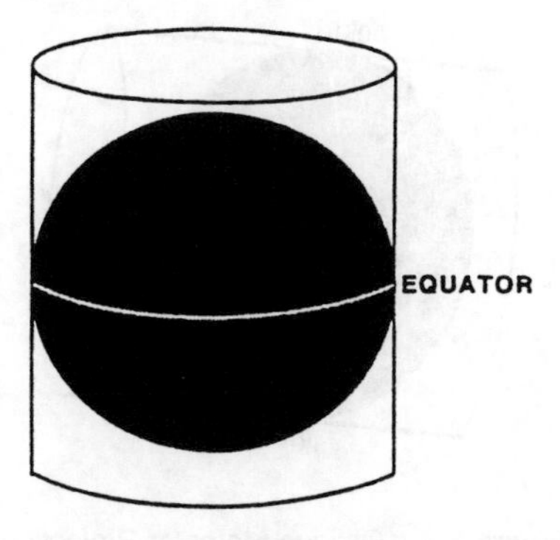

Figure 5.1. Mercator's projection.

transverse Mercator projection conforms to the curvature of the earth in a north-south direction and is therefore more accurate in that direction (Figure 5.2). Therefore, the transverse Mercator projection would be preferred over the Mercator projection for states with a long north-south axis such as Idaho, California, and Florida.

Space Oblique Mercator Projection

Historically, map projections have been based on static conditions. However, with orbiting satellites that continuously scan as the earth rotates, the satellite orbits required a new projection in which relative motion is considered. One such projection is the Space Oblique Mercator projection (Figure 5.3). This projection is the default standard projection that EOSAT uses in preprocessing of Landsat data.

MAP COORDINATE SYSTEMS

A map coordinate system is an X,Y system that can be used to reference the location of any point on the earth's surface. There are three common coordinate

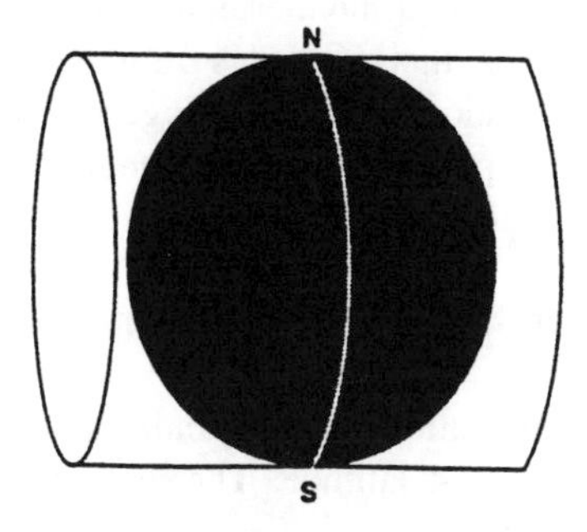

Figure 5.2. Transverse Mercator Projection.

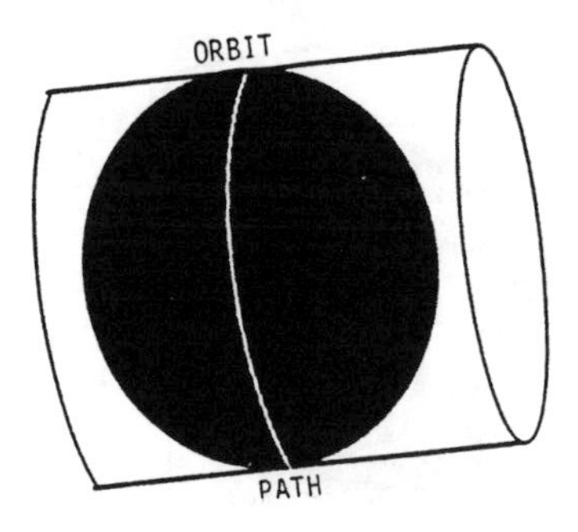

Figure 5.3. Oblique Mercator Projection.

systems used in the United States for satellite remote sensing of natural resources: geographic coordinates, Universal Transverse Mercator coordinates, and State Plane coordinates.

Geographic Coordinates

Geographic coordinates are expressed as *spherical* or *polar* coordinates of longitude and latitude. Latitude is expressed as the angular distance north or south from the equator and varies from 0 degrees at the equator to 90 degrees at the poles. Since the earth has more curvature near the equator, a degree of latitude increases in length towards the poles. For example, 1 degree of latitude originating from the equator is about 110.6 km, while 1 degree of latitude originating from the north pole is about 111.7 km. Think of the units of geographic coordinates as being analogous to time: a degree is analogous to one hour, 1 degree of latitude is equivalent to 60 minutes and 1 minute is equivalent to 60 seconds of latitude. Therefore any point on the earth's surface can be expressed as distance of latitude in degrees, minutes, seconds, and direction from the equator. For example, the University of Alaska, Fairbanks has a latitude of 64 degrees 1 minute and 30 seconds north of the equator. Longitude is expressed as the angular distance east or west of the Prime Meridian. For example, the University of Alaska, Fairbanks is 147 degrees 50 minutes 0 seconds west of the prime meridian. Where is the Prime Meridian? The Prime Meridian is an arc that originates from the poles and passes through Greenwich (southeast London), England (Figure 5.4). Since all longitude meridians converge at the poles, a degree of longitude varies from 111.3 km at the equator to 0 km at the poles.

Universal Transverse Mercator Coordinates

Distance and area calculations are difficult using a spherical coordinate system such as longitude and latitude. Therefore flat or *planar* coordinate systems are often printed on maps. You can think of a planar coordinate system as a Cartesian or X,Y coordinate system. In the United States, the X values increase toward the east and Y-values increase toward the north. The U. S. Army developed a metric planar coordinate system called the Universal Transverse Mercator (UTM) coordinate system. The system consists of zones 6 degrees wide (east-west) starting with zone 1 at 180–174° W. The zones go from 84° N to 80° S in latitude (Figure 5.5). Every zone has the same shape and size; hence the term "universal." With a separate transverse Mercator projection used for each zone, a high degree of accuracy is obtained (one part in 2,500 maximum distortion within each zone). Within each zone, there is a center meridian with an assigned false easting (or X-value) of 500,000 m. Northings (or Y-values) in the northern hemisphere are assigned as meters north of the equator. Since the zone meridan "touches" the globe, there is no scale distortion along the meridian and slight scale distortion near the east or west edge of each UTM zone.

State Plane Coordinates

In 1938, the U.S. Geologic Survey worked out a scheme of grid systems that could be fit to the irregular shapes of individual states. This system allowed for local land surveys to be connected with one another. New Jersey was the first state to legalize the use of state plane coordinates for property surveys. There are more than 130 State Plane zones, each with its own map projection and grid network. Zone boundaries follow state and county lines and each zone is relatively small, therefore distortion within each zone is less than one in 10,000. State plane coordinates are called departures (X-values) and latitudes (Y-values) and are expressed in feet.

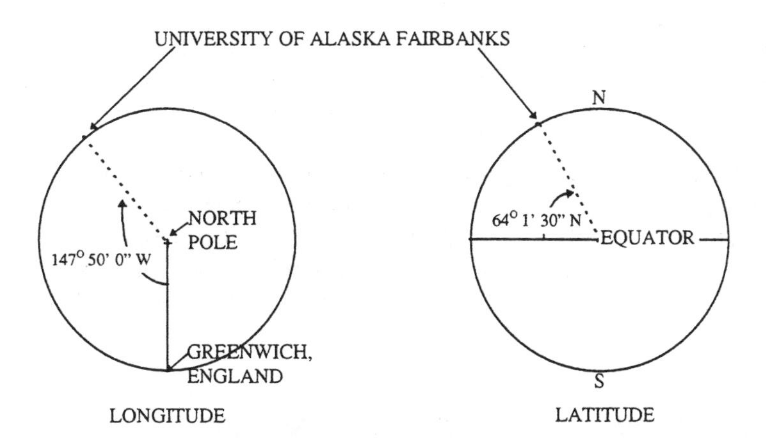

Figure 5.4. Longitude and latitude of the University of Alaska Fairbanks.

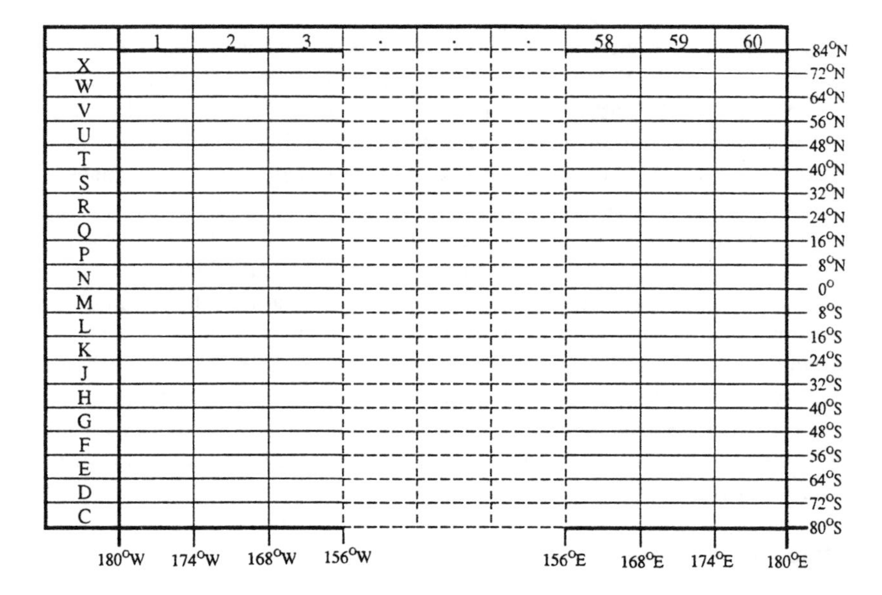

Figure 5.5. Universal Transverse Mercator grid zones. Each zone is 6 degrees wide and has its own transverse Mercator projection.

Map Coordinates on USGS 7.5-Minute Maps

In the conterminous United States, maps produced by the U. S. Geological Survey and used in natural resource management are often from the 7.5-minute quadrangle series. A 7.5-minute quadrangle has a width of 7.5 minutes (longitude) and a length of 7.5 minutes (latitude). Geographic coordinates are delineated as interior black tics every 2.5 minutes along the borders of 7.5-minute quadrangles (Figure 5.6). These tic locations are often used in registering a map to a coordinate system for use with a digitizing table. State Plane tics are usually printed every 10,000 ground-feet as black exterior tics along the borders of USGS 7.5-minute maps. UTM tics are usually printed every 1000 ground-meters as blue exterior tics along the borders of USGS 7.5-minute quadrangles. Usually, two principal digits are printed as large numbers and the remaining digits are printed as small superscript numbers. For example, a northing of $^{51}65^{000\text{m}}$N represents 5,165,000 meters north of the equator. The last three digits (000) of a UTM coordinate are sometimes not printed with UTM tics. Therefore, a false easting of $^{5}92$ represents a false easting of 592,000 meters.

U.S. NATIONAL MAP ACCURACY STANDARDS

Usually USGS 7.5-minute maps are published with the statement "this map complies with National Map Accuracy Standards." Since these maps are

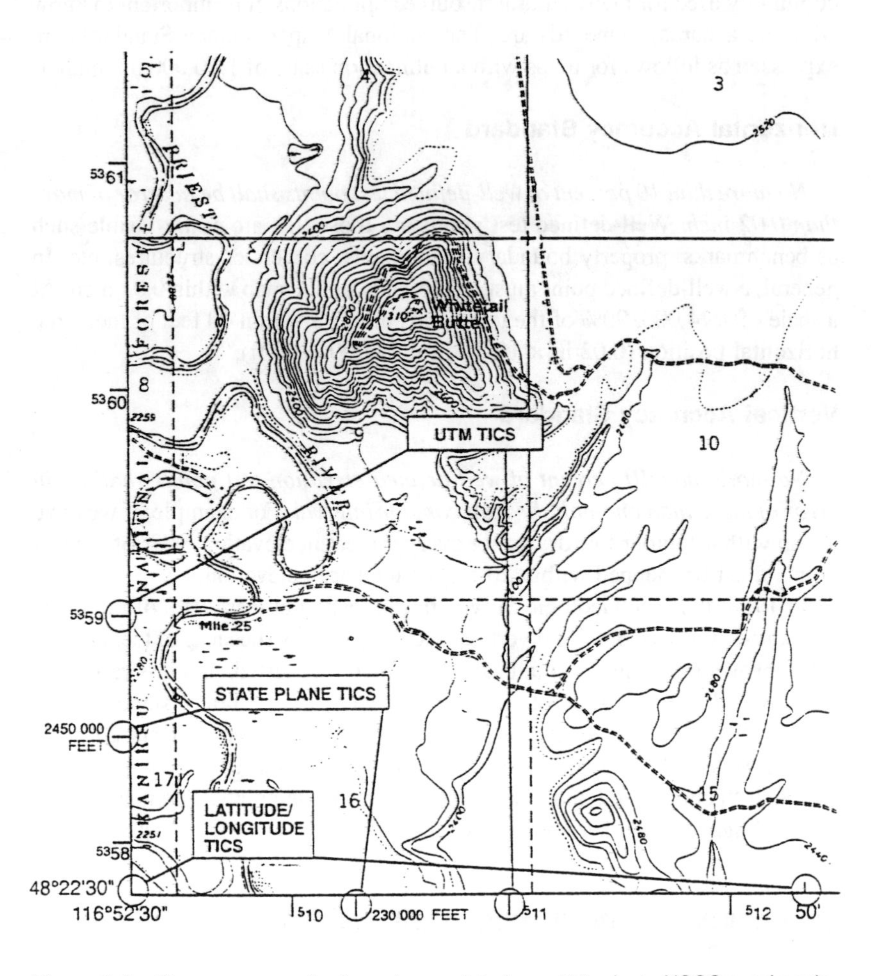

Figure 5.6. Three map-coordinate systems printed on a 7.5-minute USGS quadrangle.

U.S. NATIONAL MAP ACCURACY STANDARDS

Usually USGS 7.5-minute maps are published with the statement "this map complies with National Map Accuracy Standards." Since these maps are commonly used for many natural resource applications, it is important to know what the accuracy standards are. The National Map Accuracy Standards are expressed as follows for maps with a publication scale of 1:20,000 or smaller:

Horizontal Accuracy Standard

No more than 10 percent of well-defined test points shall be in error of more than 0.02 inch. Well-defined test points are those that are easily visible such as benchmarks, property boundary monuments, corners of structures, etc. In general, a well-defined point must be plottable on the map within 0.01 inch. At a scale of 1:24,000, 90% of the test points must be within 40 feet of their true horizontal position (0.02 in × 1ft/12 in × 24,000 = 40 ft).

Vertical Accuracy Standard

No more than 10 percent of well-defined elevation test points shall be in error of more than one-half the map contour interval. For example, if we have a map with a contour line drawn for every 40 feet in elevation, 90% of the test points must be mapped within 20 feet of their true elevation.

In 1958, the U.S. Geologic Survey began testing at least 10% of the maps it produced for horizontal and vertical accuracy. Today (because of technological improvements) only a small sample of maps are tested, and it is rare for a recently published 7.5-minute map to not meet the accuracy standards. In testing a randomly selected map, the U.S. Geologic Survey selects 20 well-defined points. Field teams then use sophisticated surveying equipment to establish the horizontal and vertical position of these points. The map is then checked against field survey results.

SELECTION OF GROUND CONTROL POINTS

Ground Control Points From Maps

The first step in the image rectification process is the selection of ground control points. Ground control points (GCPs) are locations that can be delineated on a satellite image and on a corresponding map coordinate system. Typically, these locations are easily identified pixels (ground control pixels), such as the corner of a woodlot, the intersection of roads, and rock outcrops (Figure 5.7). Since the ground control points will be used to fit a rectification model, the more high-quality points selected the better. However, accurate selection of ground control points usually becomes progressively more difficult after the most obvious points are selected. It is important that selected ground control points are well distributed across the area. For example, suppose we selected all our ground control points from the agricultural areas in a valley and then used these points to develop a rectification model. The resulting rectified image would probably have acceptable positional accuracy in the agricultural areas, but may have significant distortions in the mountainous forest areas.

Map coordinates of ground control points are often determined from U. S. Geological Survey 7.5-minute maps by using a digitizing tablet (Figure 5.7). An alternative field method for obtaining coordinates of locations is to use a global positioning system receiver.

Ground Control Points Using GPS Receivers

Global Positioning System (GPS) is a satellite-based navigation system managed by the U.S. Department of Defense. The system is based on precise timing of radio signals from at least 3 satellites. With one GPS receiver, it is

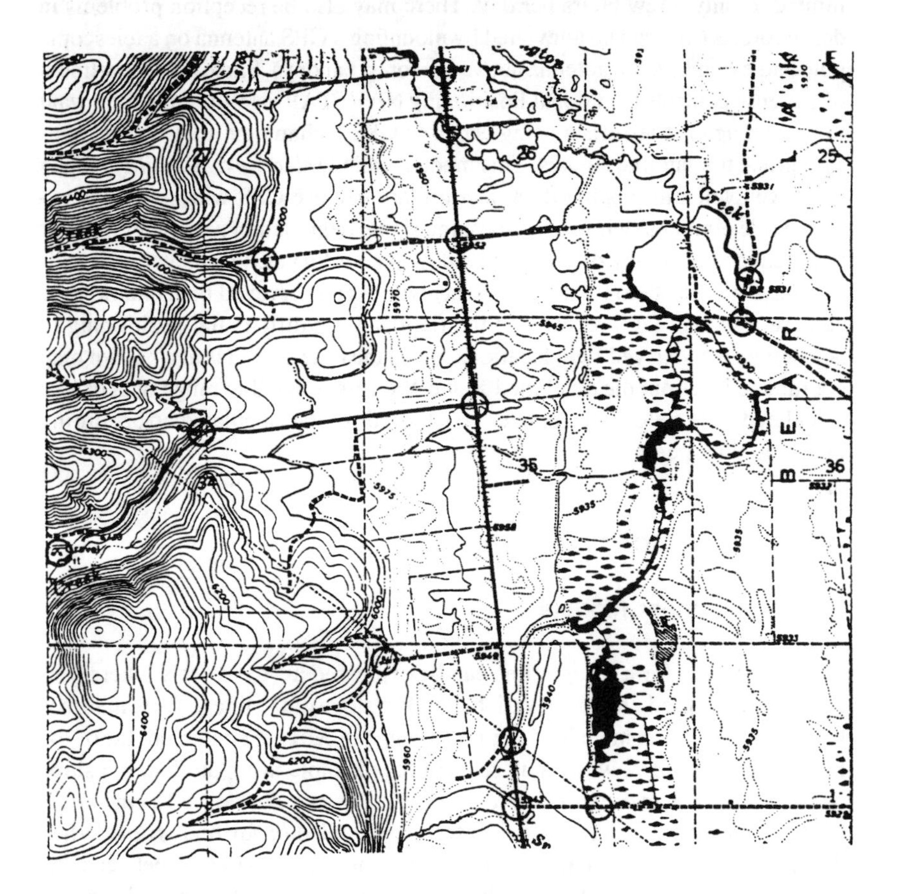

Figure 5.7. Typical ground control points (GCPs) obtained from a USGS 7.5-minute quadrangle.

possible to obtain map coordinates of almost any location in the field with a horizontal positional accuracy of 15 to 25 meters. If a GPS receiver is set up over a surveyed benchmark as a base station and another GPS receiver is used in the field, then a horizontal accuracy of 2 to 5 meters is possible through a method called *differential positioning*. In fact, surveyors use very expensive GPS receivers and differential positioning to record locations with an accuracy tolerance of less than 1 cm!

There are several limitations of using GPS in determining ground control location map coordinates. Since the technology is still fairly new, the cost of GPS receivers can be in the thousands of dollars — ranging from around $1,000 for simple, portable field-grade receivers to $75,000 for surveying grade receivers. Since at least 3 satellite signals must be received to estimate an X,Y position, a clear view toward the satellites above the horizon is crucial. Therefore, GPS reception of satellite radio signals in narrow canyons might be limited to only a few hours per day. There may also be reception problems in dense forests (this can be alleviated by mounting a GPS antenna on a telescopic pole and extending the antenna above the forest canopy). Since the Department of Defense controls the Global Positioning NAVSTAR satellites and they want to protect against terrorists using accurate GPS technology, they select times to degrade the satellite signals. This mode is called *selective availability*. When selective availability is in effect, the military can use GPS with normal accuracy while civilians (who do not have the signal degradation parameters) suffer with reduced accuracy of GPS-derived map coordinates. Fortunately, if civilians have a base station with a GPS receiver over a surveyed benchmark, they can use a second GPS receiver in the field and use differential positioning to adjust the GPS-derived field locations. Since the base station is at a known map coordinate location, the combined X and Y errors in satellite range data can be calculated. The field data can later be adjusted for the calculated errors during the same time period.

IMAGE RECTIFICATION MODELS

Review of Linear Models

The equation for a straight line can be expressed as $Y = a + bX$ where Y is the value to be predicted from variable X. The slope of the line (b) is sometimes termed rise/run or change in Y per unit change in X. The Y intercept (a) of the line is the value of X when Y is zero. In the example in Figure 5.8, the slope of the line is 0.8 and the Y-intercept is 3.6. We can develop a linear model given X, Y data by using a statistical technique called *linear regression*. When developing a linear model, there is usually some residual error between the predicted and actual Y values. With linear regression, the total squared residual error is minimized. In Figure 5.8, this total squared residual error is 3.6, and this is the best possible line in terms of minimizing this error.

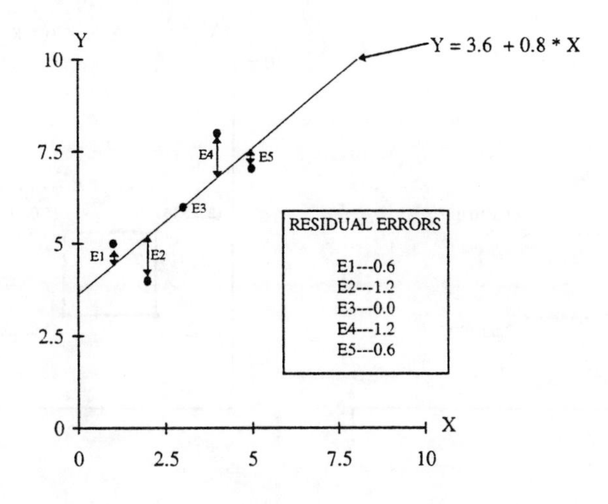

Figure 5.8. Linear regression model from 5 X,Y measurements.

Affine Coordinate Transformations

The *affine coordinate transformation* models are commonly used for rectification of digital satellite images. Scale, translation, rotation, and/or skew distortions can be modeled using these linear models. Think of the transformations as changing the x and y axes to make the transformations (Figure 5.9).

The affine transformations can be expressed as:

$$X' = a_0 + b_1 X + b_2 Y \qquad Y' = a_1 + b_3 X + b_4 Y$$

Where X,Y are old image coordinates and X′,Y′ are new, rectified image map coordinates. Typically, for each selected ground control point, the X and Y are the pixel column and row numbers and X′, Y′ are the map coordinates. Least squares regression is then used to develop the affine transformation coefficients (a_0, a_1, b_1, b_2, b_3, and b_4.). The b_1 and b_4 coefficients are X and Y scaling coefficients. The constants a_0 and a_1 are X and Y translation terms. The b_2 and b_3 coefficients are X and Y rotation or skewness coefficients. Therefore, assuming that the image is from a relatively flat area, geometric corrections for scale, translation, rotation, and/or skew is possible using the affine transformations (Figure 5.9).

Polynomial Models

The affine transformation is a first-order polynomial (since the predictor variables are raised to the power of 1). A higher order transformation may be needed for images from areas of rugged terrain, or when severe panoramic distortion exists (Table 5.1). The required minimum number of ground control points increases as the polynomial order increases. For example, you need two

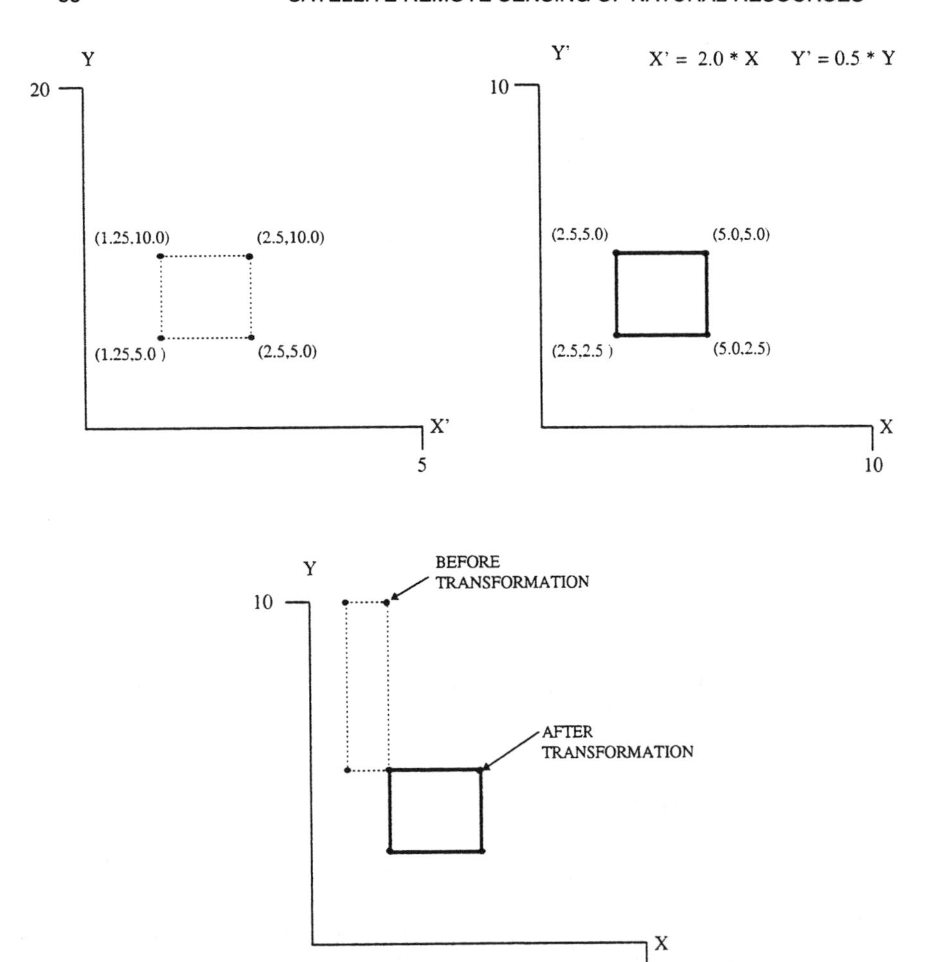

Figure 5.9A. Possible affine transformation: scaling. Both axes and origin are fixed, X and Y scale changes.

points to fit a straight line $Y = a_0 + b_1X$, three points to fit a plane ($Z = a_0 + b_1X + b_2Y$), and so on. Likewise, as polynomial order increases, there are more model coefficients, and therefore more required minimum number of ground control points.

Higher order polynomials are usually required for correction of panoramic distortion. Panoramic distortion occurs in satellite imagery due to a wide scan angle (such as AVHRR images), or off-nadir viewing (such as in SPOT HRV images acquired using a pointable mirror for oblique imaging). Think of this analogy: panoramic distortion is a major source of distortion in oblique aerial photographs — therefore, it is much easier to develop a planimetric map from vertical aerial photography. Likewise, second or third-order polynomials are required to rectify complex distortions such as panoramic distortion, while

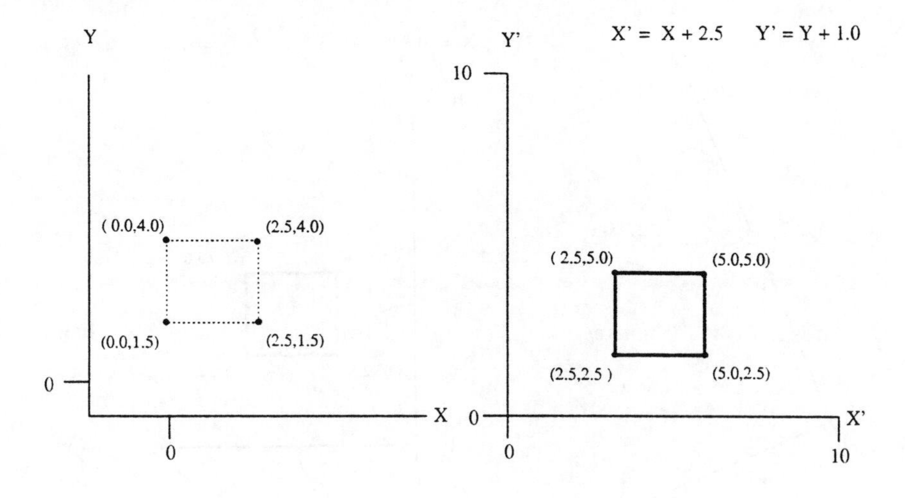

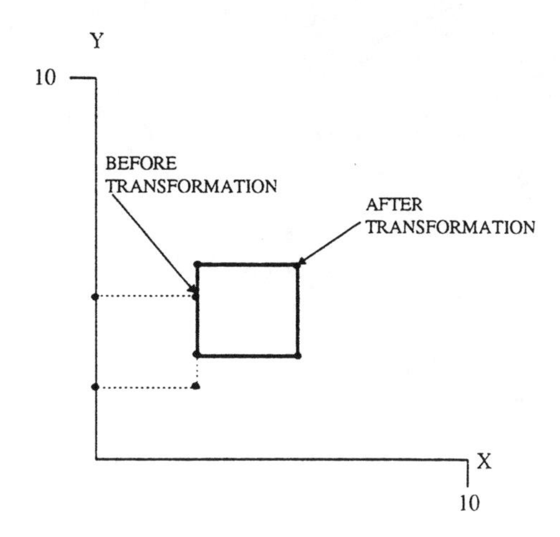

Figure 5.9B. Possible affine transformation: translation. X and Y origin is moved.

simple affine transformations can adequately rectify digital images that have little panoramic distortion and are from relatively flat terrain.

How Many GCPs?

Since we are attempting to develop statistical rectification models, a "large" sample size of ground-control points is desired. Ideally, the larger the sample size, the better our model will fit, assuming that we have "many" high-quality ground control pixels. How large? This depends on the terrain, the size of the image area, and the availability of good control pixels. For example, if we are going to rectify an image from a relatively flat landscape with a regular grid of high-contrast roads (good ground-control pixels), we might require only a few ground-control points. Welch et al. (1985) showed that 3 to 5 ground

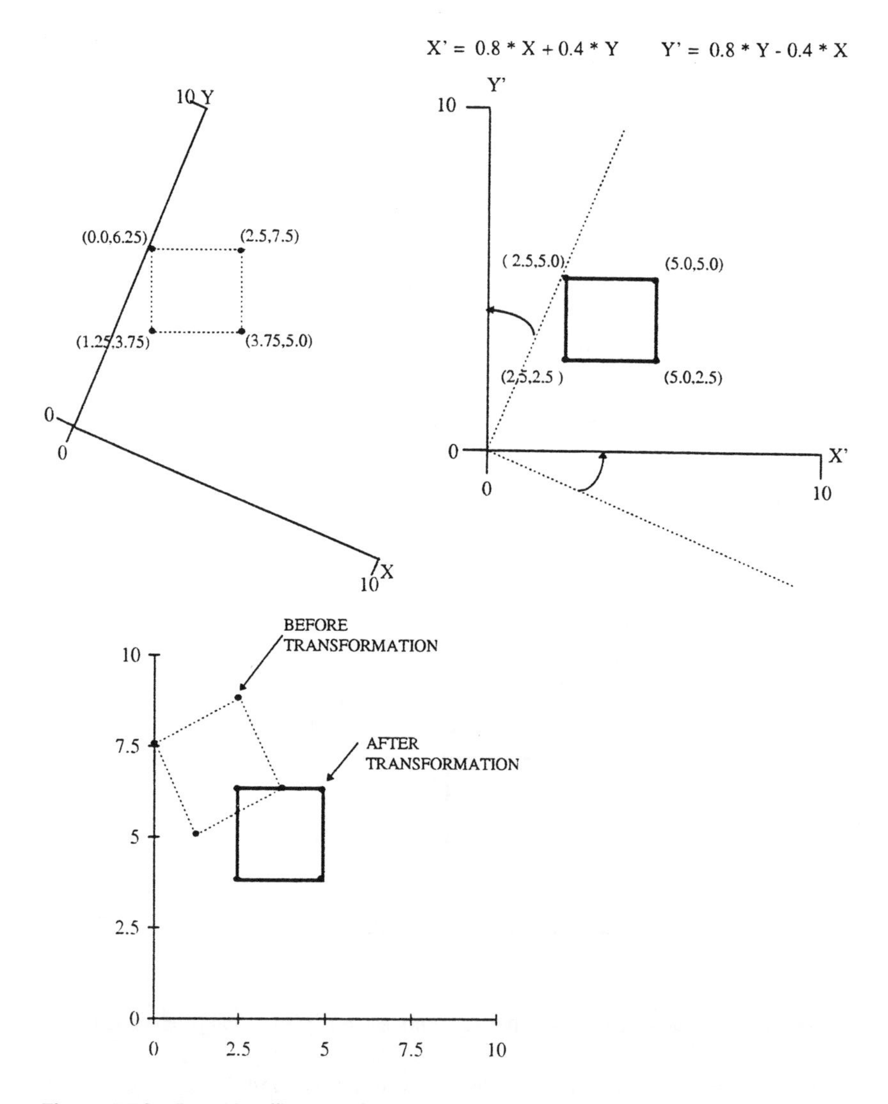

Figure 5.9C. Possible affine transformation: rotation. X and Y axes are rotated from the origin.

control-points could be used to develop acceptable transformations for a Landsat Thematic Mapper quarter scene from Iowa. However, in other areas many more GCPs will be needed to develop satisfactory rectification models. For example, the UK National Remote Sensing Center uses more than 100 ground control points to develop transformations for rectifying Landsat MSS scenes (Davison 1986).

Images from rugged terrain areas would require many ground control points to develop an acceptable polynomial rectification model In some areas, such as interior Alaska, images have poor-quality ground-control pixels since there are

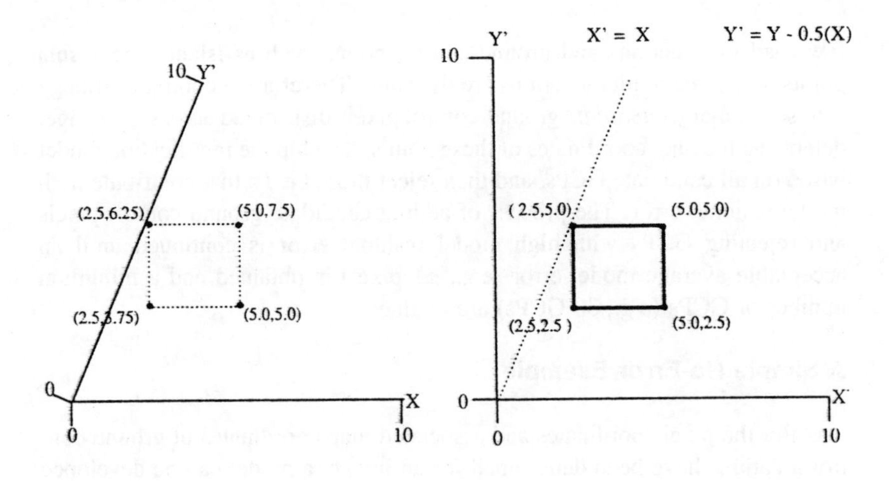

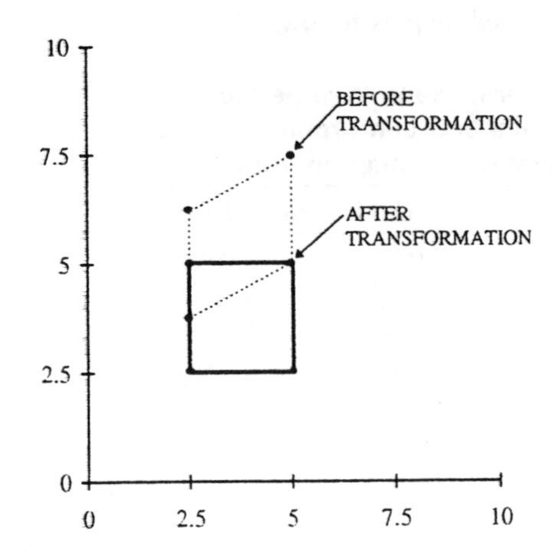

Figure 5.9D. Possible affine transformation: skewness. Rotation of a single axis.

Table 5.1.
Common polynomial transformations used for image rectification

Polynomial Order	Equation Form	Minimum Number of GCPs Required
First	$X' = a_0 + b_1X + b_2Y$	3
Second	$X' = a_0 + b_1X + b_2Y + b_3X^2 + b_4XY + b_5Y^2$	6
Third	$X' = a_0 + b_1X + b_2Y + b_3X^2 + b_4XY + b_5Y^2 + b_6X^3 + b_7X^2Y + b_8XY^2 + b_9Y^3$	10

few road intersections and ground-control points such as islands, peninsula points, and stream inlet locations are dynamic. Therefore, a common strategy is to select many *candidate* ground-control pixels distributed across the image, determine the map coordinates of these points, develop the rectification model based on all candidate GCPs, and then reject those GCPs that contribute high model residual error. The process of adding candidate ground control pixels and rejecting GCPs with high model residual error is continued until an acceptable average model error (e.g., ±1 pixel) is obtained and a minimum number of GCPs (e.g., 30 GCPs) are retained.

A Simple No-Error Example

After the pixel coordinates and associated map coordinates of ground control locations have been determined for an image, a model can be developed to geometrically correct the image so that it "fits" a map projection. Let's look at a ideal (no error) example. Imagine that we selected four ground control locations from an image and associated map as follows:

Table 5.2. Perfect (no errors) example of perfect correspondence between ground control pixel coordinates and map coordinates

Image Coordinates		UTM Coordinates	
Pixel X	Pixel Y	Easting	Northing
1	3	518,000	5,190,940
2	1	518,030	5,191,000
3	4	518,060	5,190,910
4	2	518,090	5,190,970
5	5	518,120	5,190,880

Using the ground control information from Table 5.2, we can develop a linear model to predict map UTM easting and northing from any pixel X and Y values (Figure 5.10).

Using the idealized linear rectification models, we can rectify the an image as shown in Figure 5.11.

Using our linear rectification models, fill in the following empty UTM map grid with the appropriate image pixels (Figure 5.12). Then check your work by looking at the solution in Appendix A. As an example, column 3, row 3 (X-UTM = 518 000, Y-UTM = 5 190 970) would be filled in as follows:

$$X = -17,265.666 + 1/30 * 518\ 000 = 1$$

$$Y = 173,034.333 + 1/30 * 5,190,970 = 2$$

(pixel 1,2 has a value of 77, therefore the map grid cell is filled with this value)

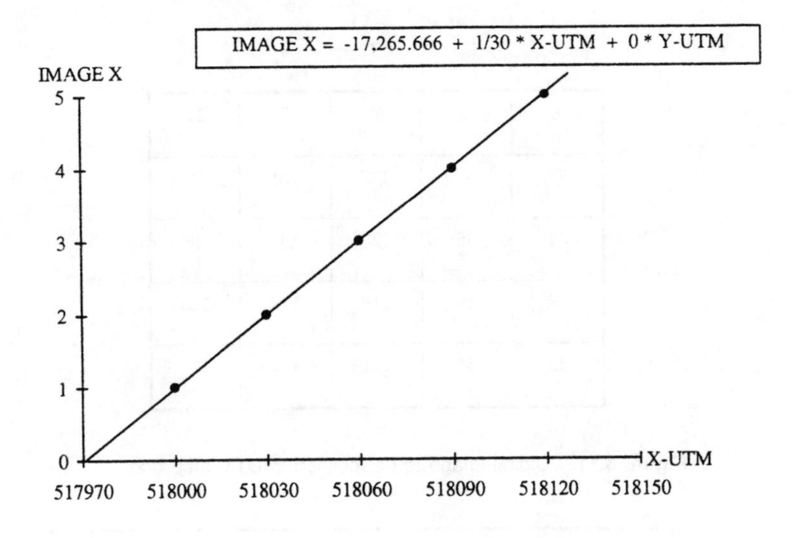

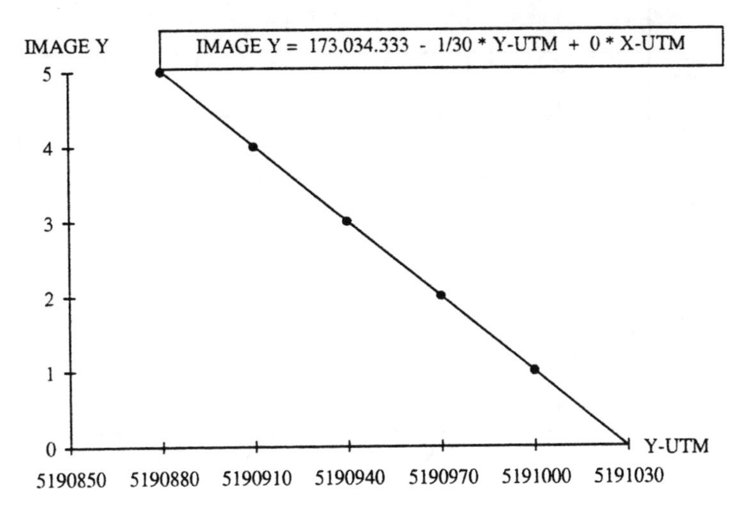

Figure 5.10. Perfect linear rectification models.

A Simple Example With Errors

Now let's consider a simple example with errors. Imagine that we want to rectify the image in Figure 5.13 using extremely accurate GPS surveyed ground-control points.

Based on the four ground-control pixels, we develop the GCP file as shown in Table 5.3.

We can use a statistical method called linear regression to develop the models to transform pixel coordinates to ground coordinates. In other words, we develop a model (or equation) to predict pixel file coordinates for locations on the map. The affine transformations are:

	1	2	3	4	5
1	6	5	63	172	17
2	77	37	127	151	39
3	61	35	120	132	38
4	59	58	119	128	113
5	55	54	101	111	112

Figure 5.11. Digital image to be rectified to UTM map grid.

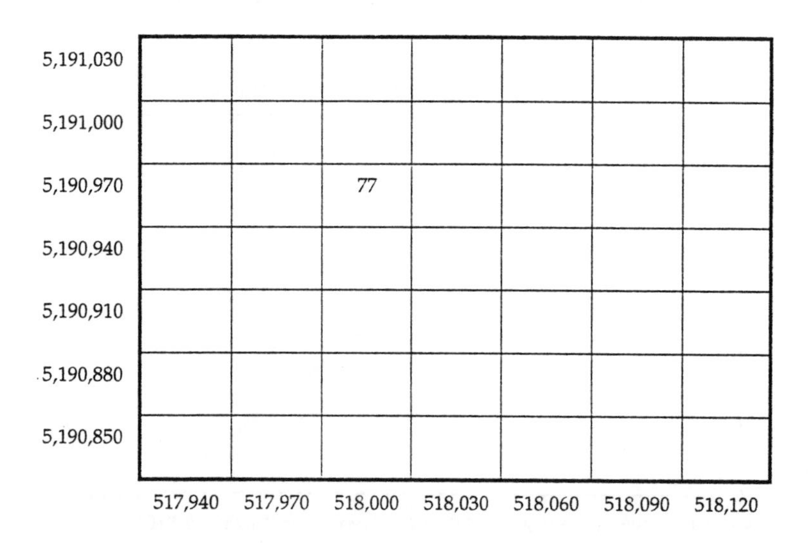

Figure 5.12. Empty UTM map grid to be filled using rectification model.

Table 5.3. Simple example of selected ground-control pixels and associated GCP map coordinates

Pixel File Coordinates		Map Coordinates	
X	Y	X	Y
1	1	31	64
1	5	10	30
3	4	32	28
5	5	44	9

$$GCPX_{map} = 1.731 + 0.0851(X_{pixel}) - 0.0526(Y_{pixel})$$

$$GPCY_{map} = 8.078 - 0.0526(X_{pixel}) - 0.0851(Y_{pixel})$$

We can compute the X and Y error of our rectification models for each ground-control location (Table 5.4).

Sometimes it is difficult to accurately locate ground-control locations on an image. Therefore some pixels may not fit our rectification models closely. By

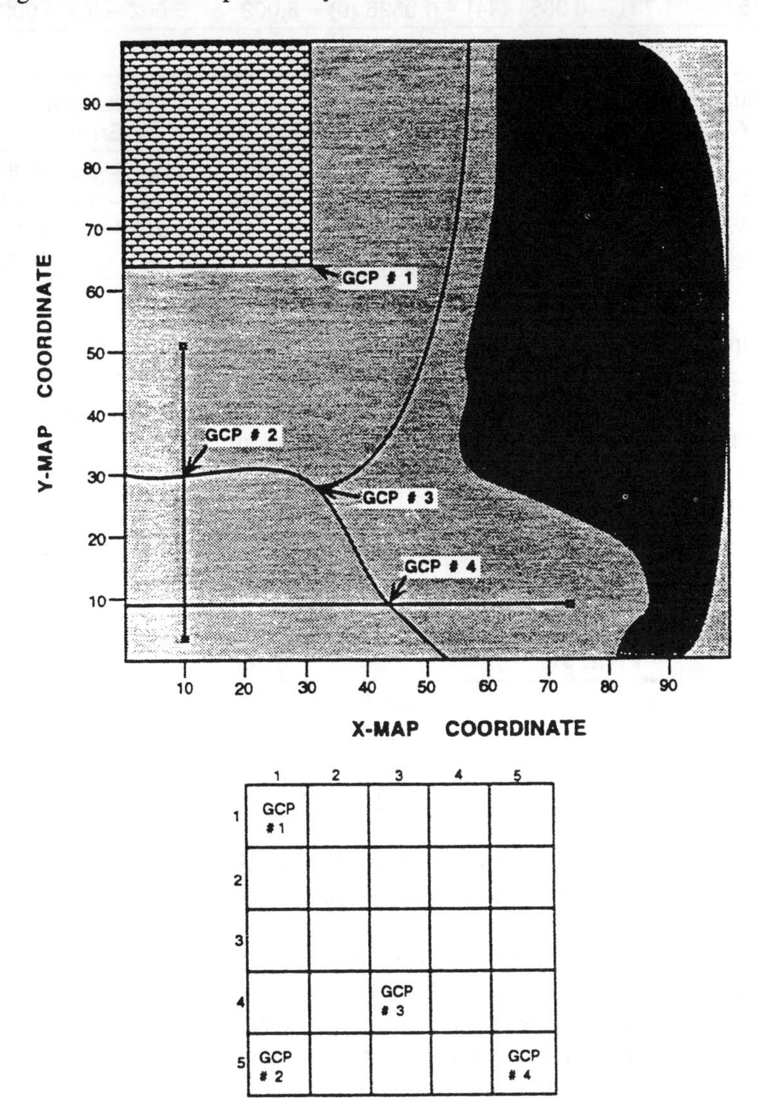

Figure 5.13. Simple example of ground-control pixels and associated map.

Table 5.4. Residual errors from simple linear rectification model

X Residual Errors

Actual X	Predicted X	Residual Error
1	1.731 + 0.0851 (31) – 0.0526 (64) = 1.003	1.003 – 1 = 0.003
1	1.731 + 0.0851 (10) – 0.0526 (30) = 1.004	1.004 – 1 = 0.004
3	1.731 + 0.0851 (32) – 0.0526 (28) = 1.004	2.980 – 3 = –0.020
5	1.731 + 0.0851 (44) – 0.0526 (9) = 5.002	5.002 – 5 = 0.002

Y Residual Errors

Actual Y	Predicted Y	Residual Error
1	8.078 – 0.0526 (31) – 0.0851 (64) = 1.001	1.001 – 1 = –0.001
5	8.078 – 0.0526 (10) – 0.0851 (30) = 4.999	4.999 – 5 = –0.001
4	8.078 – 0.0526 (32) – 0.0851 (28) = 4.012	4.012 – 4 = 0.012
5	8.078 – 0.0526 (44) – 0.0851 (9) = 4.998	4.998 – 5 = –0.002

computing the error of each ground-control location, we can assess which ground-control pixels do not accurately fit our rectification models. This error is called the *residual or root squared error,* and is basically the straight line distance between predicted and actual pixel locations (Figure 5.14).

We can compute the root squared error to compare how well each ground-control points fit the linear model:

Table 5.5.
Error contribution by each ground-control point

Ground Control Point #	Error
1	$\text{SQRT}(0.003^2 + 0.001^2) = 0.003$
2	$\text{SQRT}(0.004^2 + 0.001^2) = 0.004$
3	$\text{SQRT}(0.020^2 + 0.012^2) = 0.023$
4	$\text{SQRT}(0.002^2 + 0.002^2) = 0.003$

In this example, point 3 had the poorest fit to the model. In application, this may be due to selecting a ground-control location that was difficult to delineate on the image.

The average error of the model is often called the RMS (for root mean squared) error and is often reported as a tolerance of rectification. For example, if we rectified an image with a root mean squared error of 1, we would expect that, on the average, the accuracy of the map coordinates of our rectified image to be about ±1 pixel. In other words, an RMS of 1.0 for rectified 30-meter Landsat data would mean that, assuming the ground-control points were well

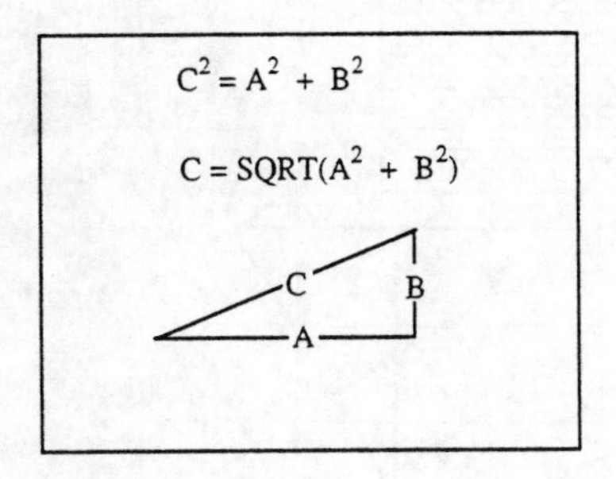

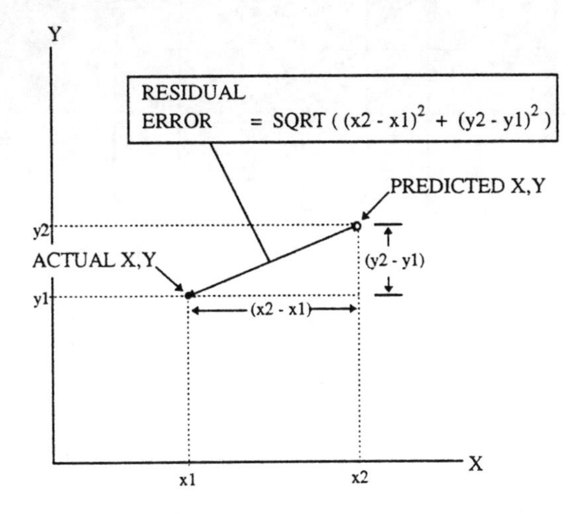

Figure 5.14. Root squared error computed as straight line distance between predicted and actual pixel locations.

distributed throughout the image, our expected average positional accuracy of the rectified pixels would be ±30 meters from their true map position.

The RMS error in our simple example can be calculated as:

$$\frac{0.003 + 0.004 + 0.023 + 0.003}{4} = 0.008$$

The linear model we developed can be portrayed graphically as shown in Figure 5.15.

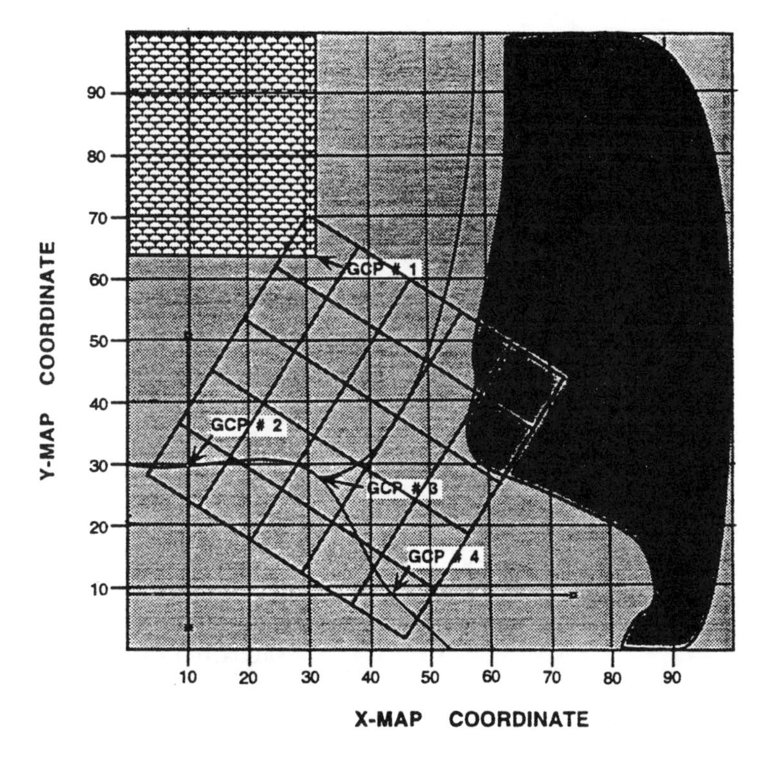

Figure 5.15. Original image oriented over planimetric map.

PIXEL RESAMPLING METHODS

You can think of image rectification as being analogous to picking up the original image, unwarping it, and then overlaying it on top of an empty map grid. Each map-grid pixel is then assigned pixel values from the image. However, the image will not perfectly match a map grid on a pixel-by-pixel basis (Figure 5.16). Therefore the original image pixels need to be resampled to fill in the new rectified (or map grid) pixels. There are three common resampling methods used to fill the empty map-grid pixels: nearest neighbor, bilinear interpolation, and cubic convolution resampling.

Nearest Neighbor Resampling

In nearest neighbor resampling, each output (rectified image) pixel is assigned the digital values of the nearest pixel from the original image. In Figure 5.16, the closest original pixel value to the location of rectified pixel (map pixel to be filled) has a digital value of 100.

This resampling method is the fastest of the three commonly used methods. It also retains the original pixel values which may be important in some quantitative spectral studies. However, this method produces jagged-looking edges, especially on linear features such as roads and shorelines which appear as staircase shapes, rather than as smooth lines.

Bilinear Interpolation Resampling

In bilinear interpolation, the digital value of each output pixel is a weighted average of the four closest original digital values (Figure 5.17).

Bilinear interpolation is interpolation in 2-dimensions. For example, imagine Figure 5.18 as an enlargement of the center of Figure 5.17. We know that the original values from the four nearest original pixels (A = 20, B = 20, C = 20, D = 100). Using linear interpolation, what would you expect the values to be at point E and F?

Interpolation in the Y-direction:

Point E: range of values from pixel A to pixel B = 20 to 100

= 100 – 20 = 80

interpolated value = 20 + 0.8(80) = 84

Point F: range of values = 20 to 20 = 20 – 20 = 0

interpolated value = 20 + 0.80(0) = 20

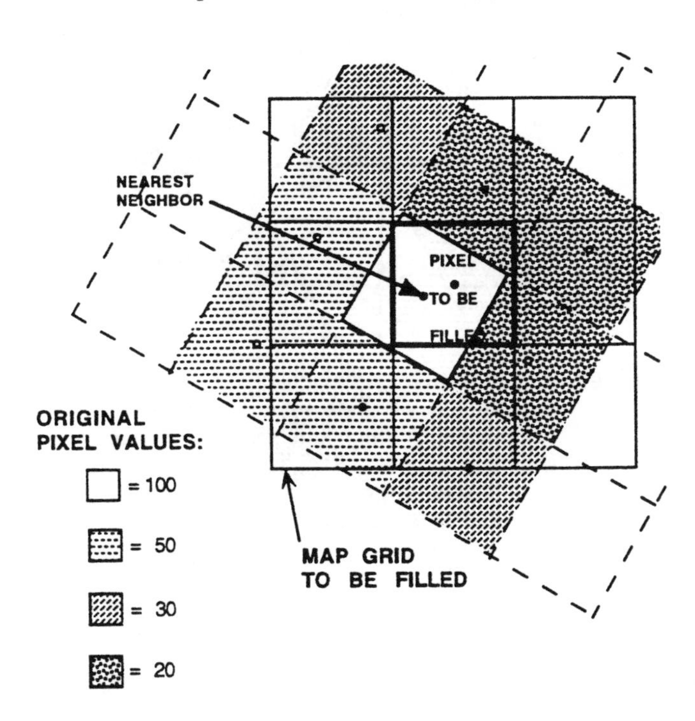

Figure 5.16. Nearest neighbor resampling to fill map grid pixel.

The final step is to interpolate in the X direction between point E (digital value of 84) and point F (digital value of 20).

Interpolation in the X-direction:

$$\text{range of values} = 20 \text{ to } 84 = 84 - 20 = 64$$

$$\text{interpolated value} = 20 + 0.85(64) = 74.4$$

Therefore, the map grid pixel will be filled with a digital value of 74.

Notice that pixel values closest to the map pixel to be filled have a greater influence in the resampled value compared with pixels that are farther away. Also notice that we obtained a much different resampled value using bilinear interpolation (74) as compared with the value we obtained using nearest neighbor resampling (100). Bilinear interpolation produces rectified images

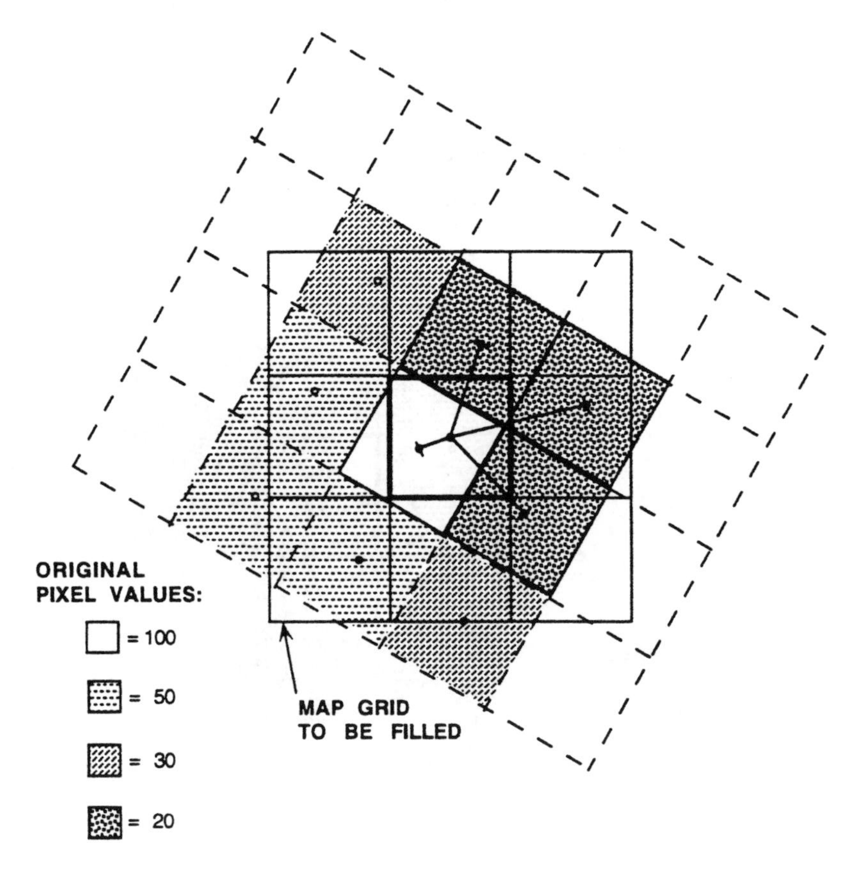

Figure 5.17. Bilinear interpolation resampling of four nearest original pixel values to fill map-grid pixel.

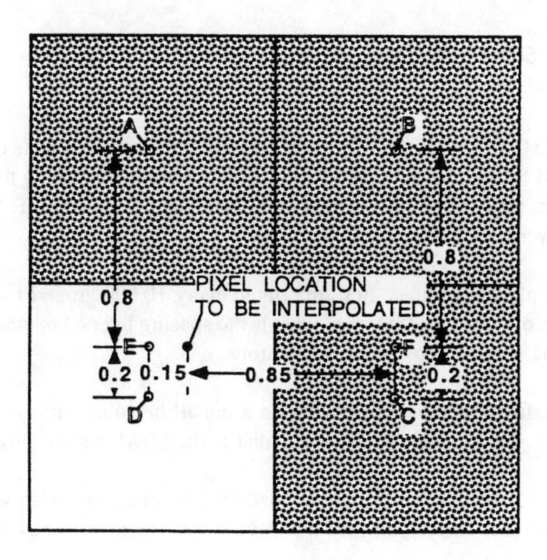

Figure 5.18. Bilinear interpolation resampling portrayed as linear interpolation in the X and Y directions.

that appear smoother than images produced using nearest neighbor resampling, especially on images that have many linear features. However, since the digital values are a weighted average of the four nearest original digital values, the method is not appropriate if one wishes to retain original digital values for analysis.

Cubic Convolution Resampling

Cubic convolution resamples pixels as a weighted average of the 16 nearest original data values; therefore it takes much longer to rectify an image with cubic convolution resampling when compared with nearest neighbor or bilinear interpolation resampling. Images resampled using cubic convolution appear smooth, but the original digital values are lost with this resampling method.

ORDERING RECTIFIED DATA

You can order Landsat and SPOT imagery as rectified or geocoded imagery. Typically the rectified imagery can be ordered as "precision corrected" or "terrain corrected." Precision corrected imagery have been rectified using ground control points. Terrain corrected imagery have been rectified using digital elevation data. The *additional* cost for a 1994 geocoded scene compared to an unrectified scene ranges from $550 for a precision-corrected SPOT scene to $1,550 for a terrain-corrected Landsat scene.

CHAPTER 5 PROBLEMS

1) Assume that you have a 7.5 minute quadrangle printed at a scale of 1:24,000. If UTM tic marks are printed every 1000 ground-meters along the border of the map, what is the spacing interval in map-inches for these UTM tic marks? Show your calculations.

2) If state plane coordinate tics are printed every 10,000 ground-feet along the borders of your 1:24,000 map, what is the spacing interval in map inches for these tic marks? Show your calculations.

3) A waterfowl biologist is working on a pintail breeding survey at longitude 159° 59′ 02″ latitude 64° 17′ 19″, what is the UTM zone of this location?

4) Why are most USGS 7.5-minute quadrangles not square? Can you think of any examples where you would expect a quadrangle to be square?

5) Given the following information about ground control points, plot the relationship between X map coordinates (the x-axis of your plot) and X pixel file coordinates (the y-axis of your plot). Then plot the relationship between Y map coordinates (the x-axis of your plot) and Y file coordinates (the y-axis of your plot).

File Coordinates		Map Coordinates	
X	Y	X	Y
1	1	10	15
4	2	13	16
3	3	12	17
1	4	10	18

From your plot, determine the coefficients (y-intercept and slope) of the linear models:

X file coordinate = $a_1 + b_1$(x ground coordinate)
Y file coordinate = $a_2 + b_2$(y ground coordinate)

a_1 = ____ a_2 = ____
b_1 = ____ b_2 = ____

6) Some people recommend rectifying an image after it has been classified. Their rational usually is: 1) Rectifying is quicker since each pixel contains

only one class value instead of many spectral values, and 2) Some spectral integrity is lost during the pixel resampling process — an unrectified image is spectrally more correct than a rectified image and is therefore preferred for image classification. List at least two reasons why you might want to rectify an image before classifying it for vegetation mapping.

Reason #1:

Reason #2:

7) If you are developing linear affine transformations of the form: $X' = a_0 + b_1 X + b_2$, $Y' = a_1 + b_3X + b_4Y$ you will always get model RMS error of 0.0 if you develop the model with 3 ground control points. Why?

8) The following transformation models were developed using a Landsat Thematic Mapper image of coastal area. After throwing out the poorest ground control pixels, seven ground control pixels remained.

GCP#	X pixel	X UTM	Y pixel	Y UTM
1	191	598,285	180	3,627,280
2	98	595,650	179	3,627,730
3	137	596,250	293	3,624,380
4	318	602,200	115	3,628,530
5	248	600,350	83	3,629,730
6	255	600,440	113	3,628,860
7	272	600,540	196	3,626,450

The transformation models are:

$$\text{X pixel} = -382.1 + 0.0341877\ (\text{Xmap}) - 0.0054810\ (\text{Ymap})$$

$$\text{Y pixel} = 130{,}163 - 0.00557618\ (\text{Xmap}) - 0.0349150\ (\text{Ymap})$$

Suppose you use these models to rectify the Landsat image to 30-meter grid cells for on-screen digitizing in a Geographic Information System. What is the expected positional accuracy (in meters) of the rectified image. Show your work!

9) Go to the library and examine some digital remote sensing studies. Find two studies that applied geometric corrections and two studies that did not. Discuss why or why not geometric corrections were used in these applications.

10) Find one study that used each of the three most popular resampling methods: nearest neighbor, bilinear interpolation, and cubic convolution. For each

study, discuss whether you feel the most appropriate resampling method was used and why?

11) Read the paper by Slonecker and Hewitt (1991). In this paper several methods of determining ground control locations are compared. Based on this one study, rank the following methods for accuracy of ground control (1 = highest, 4 = lowest)

____ geocoded image data
____ global positioning systems (GPS)
____ map interpolation using an engineers scale
____ map interpolation using a digitizing tablet

12) Suppose you were attempting to use the junction of streams within the Salmon River canyons as ground control locations. Which one of the following methods would you use and why?

____ geocoded image data
____ global positioning systems (GPS)
____ map interpolation using an engineers scale
____ map interpolation using a digitizing tablet

ADDITIONAL READINGS

Map Accuracy

Junger, S. 1991. The last mapmakers. *American Heritage.* Sept:94–98.

Rosenfield, G. H. 1971. Horizontal accuracy of topographic maps. *Surveying and Mapping.* 31:60–64.

Slonecker, E. T. and Hewitt, M. J. III. 1991. Evaluating Locational Point Accuracy in a GIS Environment. *GeoInfo Systems.* 1(8):36–44.

Thompson, M. M. and G. H. Rosenfield. 1971. On map accuracy specifications. *Surveying and Mapping.* 31:57–60.

Map Projections and Coordinate Systems

Colvocoresses, A. P. 1974. Space Oblique Mercator: a new map projection of the earth. *Photogrammetric Engineering and Remote Sensing.* 40:921–926.

Grubb T. G. and W. L. Eakle. 1988. Recording wildlife locations with the Universal Transverse Mercator (UTM) Grid System. *USDA Forest Service Research Note* RM-483. 3 pp.

Snyder, J. P. 1978. The space oblique mercator projection. *Photogrammetric Engineering and Remote Sensing.* 44:585–596.

Snyder, J. P. 1981. Map projections for satellite tracking. *Photogrammetric Engineering and Remote Sensing.* 47:205–213.

Snyder, J. P. 1982. Map projections used by the U.S. Geological Survey. *USGS Bulletin 1532.* 313 pp.

Stem, J. E. 1990. State Plane Coordinate System of 1983. *NOAA Manual NOS NGS 5.* 119 pp.

United States Army. 1973. Universal Transverse Mercator Grid. *Technical Manual TM5–241–8.* 49 pp.

Global Positioning Systems

Bertram, T. E. and A. E. Cook. 1993. Satellite imagery and GPS-aided ecology. *GPS World.* 4(10):48–53.

Corcoran, W. 1991. Six Examples of How GPS Fits into the Surveyors Tool Box. *Professional Surveyor.* 11(4):10–12.

Evans, D. L. 1992. Using GPS to evaluate aerial video missions. *GPS World.* 3:24–29.

Gibbons, G. 1992. The global positioning system as a complementary tool for remote sensing and other applications. *Photogrammetric Engineering and Remote Sensing.* 58:1255–1257.

Henstridge, F. 1991. Getting Started in GPS. *Professional Surveyor.* 11(4):4–9.

Hurn, J. 1989. *GPS: A Guide to the Next Utility.* Trimble Navigation Ltd., Sunnyvale, CA. 76 pp.

Long, D. S., DeGloria, S. D. and J. M. Galbraith. 1991. Use of the global positioning system in soil survey. *Journal of Soil and Water Conservation.* 46:293–297.

Perry, E. M. 1992. Using GPS in agricultural remote sensing. *GPS World.* 3:30–39.

Slonecker, E. T., Owecke, J. W., Mata, L. and L. T. Fisher. 1992. GPS: Great gains in the great outdoors. *GPS World.* 3:24–34.

Wilke, D. S. 1989. Performance of a backpack GPS in a tropical rain forest. *Photogrammetric Engineering and Remote Sensing.* 55:1747–1749.

Wilke, D. S. 1990. GPS location data: an aide to satellite image analysis of poorly mapped regions. *International Journal of Remote Sensing.* 11:653–638

Rectification of Digital Images

Benny, A. H. 1981. Automatic relocation of ground control points in Landsat imagery. *International Journal of Remote Sensing.* 4:337–342.

Borgeson, W. T., Batson, R. M., and H. H. Kieffer. 1985. Geometric Accuracy of Landsat-4 and Landsat-5 Thematic Mapper Images. *Photogrammetric Engineering and Remote Sensing.* 51(12):1893–1898.

Brivio, P. A., Ventura, A. D., Rampini, A. and R. Schettini. Automatic selection of control points from shadow structures. *International Journal of Remote Sensing.* 13:1853–1860.

Davison, G. J. 1986. Ground control pointing and geometric transformation of satellite imagery. *International Journal of Remote Sensing.* 7:65–74.

Ford, G. E. and C. I. Zanelli. 1985. Analysis and quantification of errors in the geometric correction of satellite images. *Photogrammetric Engineering and Remote Sensing.* 51:1725–1734.

Irish, R. 1990. Geocoding Satellite Imagery for GIS Use. *GIS World.* 3(4):59–62.

Kirby, M. and D. Steiner. 1978. Affine transformation in the solution of the geometric base problem in Landsat data. *Canadian Journal of Remote Sensing.* 4:32–36.

Labovitz, M. L. and J. W. Marvin. 1986. Precision in geodetic correction of TM data as a function of the number, spatial distribution, and sucess in matching of control points: a simulation. *Remote Sensing of Environment.* 20:237–252.

Ton, J. and A. K. Jain. 1989. Registering Landsat images by point matching. *IEEE Transactions on Geoscience and Remote Sensing.* 27:642–648.

Trinder, J. C. 1989. Precision of digital target location. *Photogrammetric Engineering and Remote Sensing.* 55(6):883–886.

Walker, R. E., Zobrist, A. L., Bryant, N. A., Gohkman, B., Friedman, S. Z. and T. L. Logan. 1984. An analyis of Landsat-4 Thematic Mapper geometric properties. *IEEE Transactions on Geoscience and Remote Sensing.* GE-22:288–293.

Welsh, R., Jordan, R. and M. Ehlers. 1985. Comparative Evaluation of the Geodetic Accuracy and Cartographic Potential of Landsat-4 and Landsat-5 Thematic Mapper Image Data. *Photogrammetric Engineering and Remote Sensing.* 51(11):1799–1812, 51:1249–1262.

Welsh, R., and E. L. Usery. 1984. Carotgraphic accuracy of Landsat-4 MSS and TM data. *IEEE Transactions on Geoscience and Remote Sensing.* GE-22:281–287.

Westin, T. 1990. Precision rectification of SPOT imagery. *Photogrammetric Engineering and Remote Sensing.* 56:247–253.

CHAPTER 6

Unsupervised Classification

INTRODUCTION

Unsupervised classification is a process of grouping pixels that have similar spectral values. Each group of similar pixels is typically called a spectral class. These spectral classes are assumed to correspond to cover type classes such as range type, stand density, timber size classes, and wetland types. Unsupervised classification procedures generally require no knowledge of existing cover types prior to classification. This may be an advantage where the image analyst does not know all the cover types in some areas. For example, you might want to classify an area that is predominantly black spruce, white spruce, balsam poplar, and aspen/birch. However, you do not know that larch stands also occur in the area. With unsupervised classification, if larch is spectrally distinct, several spectral classes corresponding to larch may be delineated. Another potential advantage of unsupervised classification is the identification of areas that have unusual spectral values; these areas may be relatively rare habitat areas within the image area.

There are many unsupervised procedures; in this chapter we will cover three general approaches: histogram-based unsupervised classification, sequential clustering and ISODATA clustering.

HISTOGRAM-BASED UNSUPERVISED CLASSIFICATION

Imagine that we have a near-infrared digital image from the taiga region of Alaska. The major cover types in the image are water, riparian willows (broadleaf shrubs), black spruce stands (coniferous trees), and rock outcrops. We could generate a histogram showing the number of pixels within each digital number class from 0 to 255 (Figure 6.1).

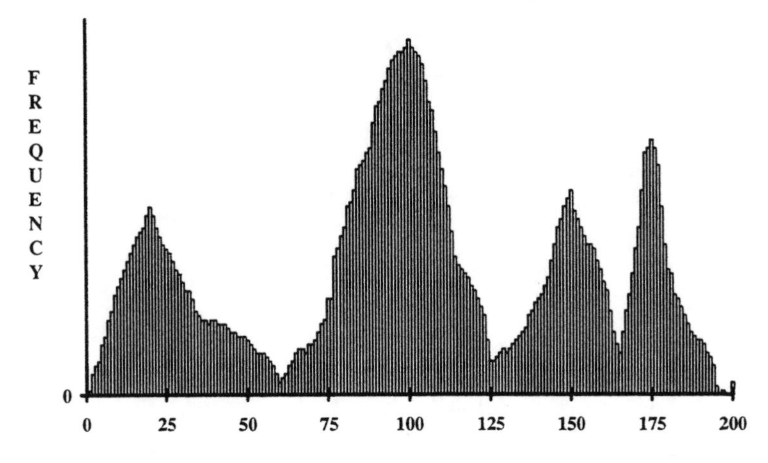

Figure 6.1. Histogram of near-infrared digital numbers from hypothetical image.

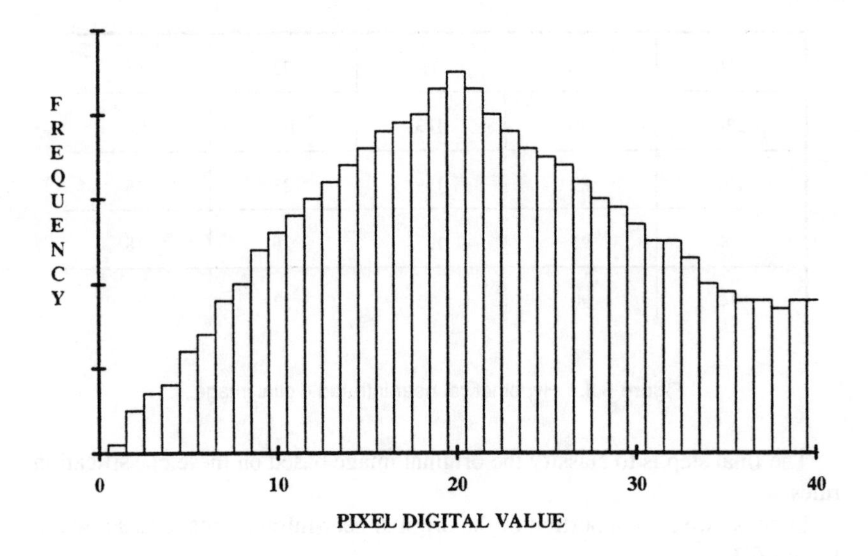

Figure 6.2. Simple hypothetical histogram of pixels from lakes and streams.

The histogram from pixels of a uniform cover type is often bell shaped. For example, open water pixels will generally have low digital values from a near infrared band. Most of the pixels might have a value of 20; however very clear deep water pixels might have values ranging down to 1, while shallow, turbid water pixels might have values in the range of 20 through 40 (Figure 6.2).

Histogram-based unsupervised classification relies on this bell-shaped assumption and follows a series of rules to delineate spectral classes. The procedure generally first determines peaks within the histogram throughout the range of image digital values. Each peak in the histogram will correspond to a spectral class. For example, in Figure 6.1 we have histogram peaks at values of 20, 100, 150, and 175. Since we have four major histogram peaks, we will delineate four spectral classes. The next step is to determine the boundaries of each spectral class. We will use a simple (but not necessarily the best) rule; the boundary between spectral classes will be half the distance between class peaks. The decision boundaries computed as half the distance between histogram peaks are 60, 125, and 162.5 (Table 6.1). Draw in these limits or decision boundaries on the histogram in Figure 6.1.

Table 6.1. Classification thresholds based on histogram peaks

Spectral Class	Range of Digital Values
1	less than 60
2	61 through 125
3	126 through 162
4	greater than 162

230	201	77	73	68
230	201	143	147	153
102	89	139	23	15
98	91	137	26	18
94	90	125	22	13

Figure 6.3. Hypothetical near-infrared digital image.

The final step is to classify the original image based on these classification rules.

Suppose we have a portion of the original near-infrared image as shown in Figure 6.3.

Using the classification rule, we can fill in our classified image by assigning spectral classes for each pixel as shown in Figure 6.4.

Finally, we could color the classified image with the following color palette (Figure 6.5):

spectral class #1 → blue
spectral class #2 → dark green
spectral class #3 → yellow green
spectral class #4 → brown

A classified image is then displayed using the color palette. Typically, your next step is to interpret what cover types correspond to each spectral class by using some tools that will be discussed later in this chapter. In this simple example, by examining the classified image, we might decide that spectral

4	4	2	2	2
4	4	3	3	3
2	2	3	1	1
2	2	3	1	1
2	2	2	1	1

Figure 6.4. Classified image values.

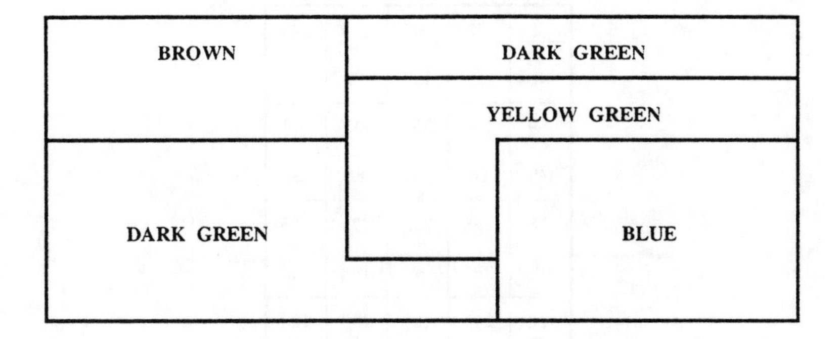

Figure 6.5. Color-coded classified image.

class #1 corresponds to water, spectral class #2 corresponds to black spruce stands, spectral class #3 corresponds to willow riparian areas, and spectral class #4 corresponds to rock outcrops.

Histogram-based unsupervised classification can be applied to multispectral images. For example, if we had a two-band image, the peaks in the histogram would be analogous to the peaks of hills or mountains (two-dimension peaks). With seven spectral bands, we would have histogram peaks in seven dimensions — we cannot draw in seven dimensions; however, the technique still works because mathematically we can calculate spectral distance in n-dimensions. You will learn more about this later in this chapter.

SEQUENTIAL CLUSTERING

Sequential clustering (also called k-means clustering) is a common strategy in unsupervised classification. The basic idea is to sequentially sample pixels and assign each sampled pixel to the nearest spectral class mean. For example, suppose we have the a two-band image as shown in Figure 6.6. One type of sequential clustering can be performed using the following clustering rules:

Pass #1

1) No more than a user-specified maximum number of spectral classes will be produced.
2) A candidate pixel will belong to the most similar spectral class in terms of spectral distance to class means.
3) If a candidate pixel is farther than a user-specified spectral distance from all existing spectral classes, then the pixel will become the first member of a new spectral class. This is the allowable radius for a spectral class; any pixel outside this radius is considered too different to be a member of this spectral class.
4) Start at pixel 1,1 and process the image sequentially from left to right. The last pixel to be processed will be the pixel in the lower right corner of the image.

50	78	88	125	244
43	65	123	99	233
59	128	209	117	67
49	98	154	88	33
78	198	205	233	205
193	231	99	198	132
22	245	109	239	100
141	241	75	202	38
58	245	14	217	114
233	249	189	156	48

Figure 6.6. Hypothetical two-band image.

Pass #2

1) Once the mean digital values for spectral classes have been determined from pass #1, classify each pixel in the image. Start at pixel 1,1 and process the image sequentially, pixel by pixel. Each pixel will be assigned a spectral class by comparing the spectral distance between each spectral class mean and the pixel being considered. The pixel will then be classified based on the minimum distance to spectral class means.

This sequential clustering process will become clearer if we graphically try the process. We will use Figures 6.6 and 6.7 with the following user-specified clustering criteria:

1) Maximum allowable number of spectral classes = 10.
2) Maximum allowable distance to spectral class mean = 40.

Let's give it a try. Starting at pixel 1,1 we have our first spectral class; this class has one member (pixel 1,1) and has mean digital values of 50 and 43. Shade in the circle representing pixel 1,1 in Figure 6.7 (use a pencil) and label this spectral class #1. Then shade in the next candidate — the symbol representing pixel 1,2 values. Is the spectral distance of this candidate close enough to become a member of spectral class #1? By looking at Figure 6.7, you can see that this distance is close enough (it's spectral distance is less than 40). Therefore pixel 1,2 is assigned as a member of spectral class #1. Spectral class #1 now has two members; thus, the class has new mean digital numbers based on the two pixels: (50 + 78)/2, (43 + 65)/2 or 64,54. Plot this new mean in Figure 6.7.

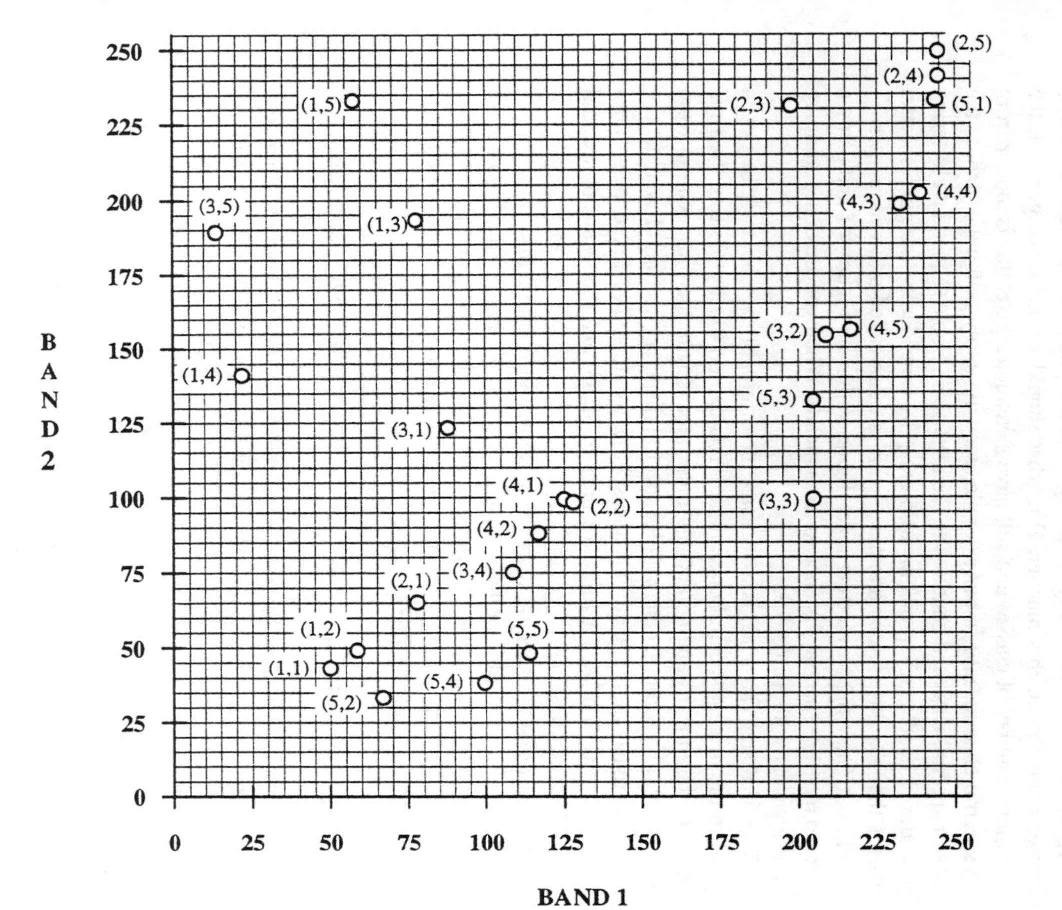

Figure 6.7. Worksheet for plotting spectral digital values of candidate pixels and spectral classes.

The next candidate pixel has digital values of 88,123; shade in this candidate. We can see that this pixel is too far from the mean digital values of spectral class #1 and therefore pixel 1,3 becomes the first member of a new spectral class: spectral class #2.

This process of sequentially assigning candidate pixels as members of spectral classes continues in this manner. But what should we do if we generate the maximum number of classes and still have a candidate pixel that is too far from all spectral class means? In our example, we will simply assign the pixel to the closest spectral class and continue to the next pixel. Some sequential clustering procedures handle this problem differently by considering merging spectral classes that are close to one another — this is often called a forced merger.

Fill out Table 6.2 by plotting pixel digital values on Figure 6.7 and using a ruler to measure spectral distance to class means. When you are done, compare your answers with Table 6.3.

The final step (pass #2) is to classify the original image by assigning each pixel to the nearest spectral class mean. You can do this by plotting the final spectral class means and comparing these means with the plot location from each pixel. The computer cannot plot spectral distances and therefore must use a formula to calculate spectral distances. How would you calculate the spectral distance between pixel 1,1 (digital values of 50,43) and the means of spectral class #1 (70.8,45.6)?

We can draw the spectral distance as shown in Figure 6.8.

Notice from this plot we have a right triangle. We know from high school trigonometry (Pythagorean theorem) that the distance c is equal to the square root of $a^2 + b^2$. Therefore the spectral distance between pixel(1,1) and spectral class #1 mean digital values is:

$$\text{Spectral Distance} = \text{SQRT}[(50- 70.8)^2 + (43 - 45.6)^2] = 20.96$$

This spectral distance is sometimes called *Euclidean distance.* Euclidean distance can easily be calculated for any number of spectral bands. For example, suppose we had a seven-band Landsat Thematic Mapper image and wanted to calculate the spectral distance between pixel 1,1 and spectral class #1:

The spectral distance in seven bands between pixel 1,1 and spectral class #1 can be computed as:

$$\begin{aligned}\text{Spectral Distance} &= \text{SQRT}[(120 - 132)^2 + (189 - 193)^2 \\ &\quad + (134 - 110)^2 + (204 - 203)^2 + (192 - 182)^2 \\ &\quad + (138 - 126)^2 + (177 - 175)^2] \\ &= 31.38\end{aligned}$$

In pass #2 of sequential clustering, the computer assigns membership to the closest spectral class means. For example, pixel 1,1 has digital values of 50,43.

Table 6.2. Worksheet for pass #1 of sequential clustering unsupervised classification

Spectral Class	Class Members (col, row)	New Class Mean
1	1,1 2,1	50.00,43.00 64.00,54.00
2	3,1	88.00,123.00
3		
4		
5		
6		
7		
8		
9		
10		

The spectral distance to each spectral class mean can be computed using the above formula. The spectral distances would be as follows:

spectral class #1 → 20.96
spectral class #2 → 80.31
spectral class #3 → 78.71
spectral class #4 → 263.70
spectral class #5 → 191.26
spectral class #6 → 152.59
spectral class #7 → 239.27

Table 6.3. Spectral class mean digital values after pass #1 of sequential clustering unsupervised classification

Spectral Class	Class Members (col, row)	New Class Mean
1	1,1	50.00,43.00
	2,1	64.00,54.00
	1,2	62.33,52.33
	5,2	63.50,47.50
	5,4	**70.80,45.60**
2	3,1	**88.00,123.00**
3	4,1	125.00,99.00
	2,2	126.50,98.50
	4,2	123.33,95.00
	3,4	119.75,90.00
	5,5	**118.60,81.60**
4	5,1	244.00,233.00
	4,3	238.50,215.50
	2,4	240.67,224,00
	4,4	240.25,218.50
	2,5	**241.20,224.60**
5	3,2	209.00,154.00
	5,3	207.00,143.00
	4,5	**210.33,147.33**
6	1,3	**78.00,193.00**
7	2,3	**198.00,231.00**
8	3,3	**205.00,99.00**
9	1,4	22.00,141.00
	3,5	**18.00,165.00**
10	1,5	**58.00,233.00**

spectral class #8 → 164.81
spectral class #9 →126.13
spectral class #10 →190.17

Therefore the pixel 1,1 (with an original digital value of 50,43) of the classified image would be assigned to spectral class #1.

We can use the plot of spectral class means (Figure 6.9) to assign the final spectral classes for each pixel. Then we can color the classified image using the color scheme given in the color palette below (Figure 6.10).

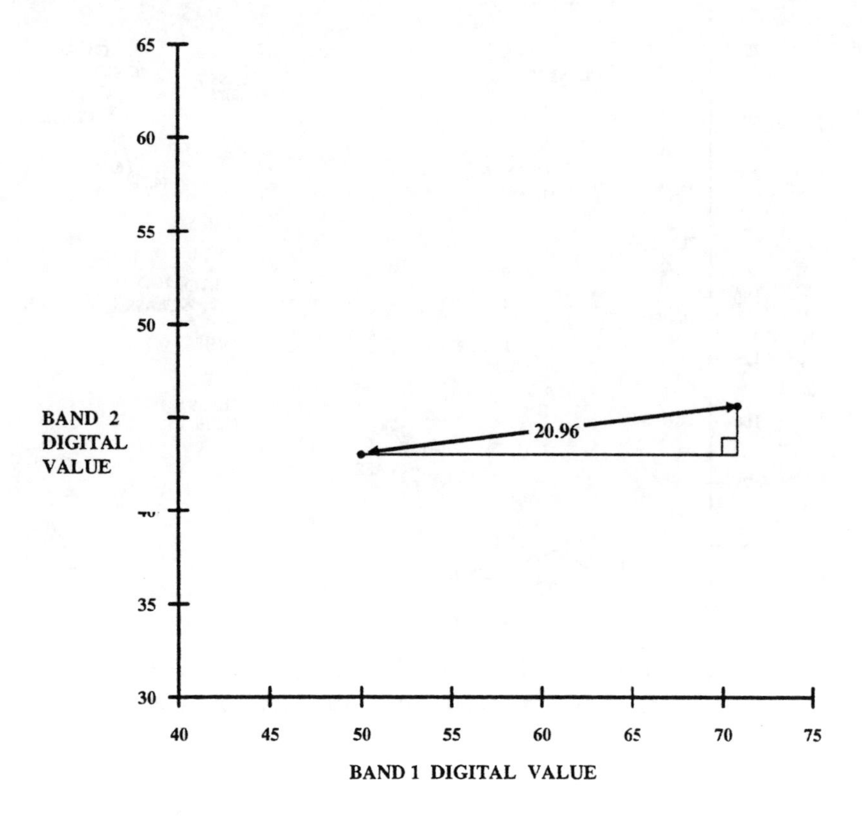

Figure 6.8. Euclidean or spectral distance using two bands.

ISODATA CLUSTERING

Another common clustering algorithm is the Iterative Self-Organizing Data Analysis or ISODATA algorithm. ISODATA uses a strategy that is opposite of sequential clustering; instead of starting with one pixel as a member of spectral class #1, ISODATA starts with all pixels as a member of spectral class

Table 6.4. Hypothetical seven-band digital values for spectral distance calculation

Spectral Band	Pixel 1,1 Digital Value	Spectral Class #1 Mean Digital Value
1	120	132
2	189	193
3	134	110
4	204	203
5	192	182
6	138	126
7	177	175

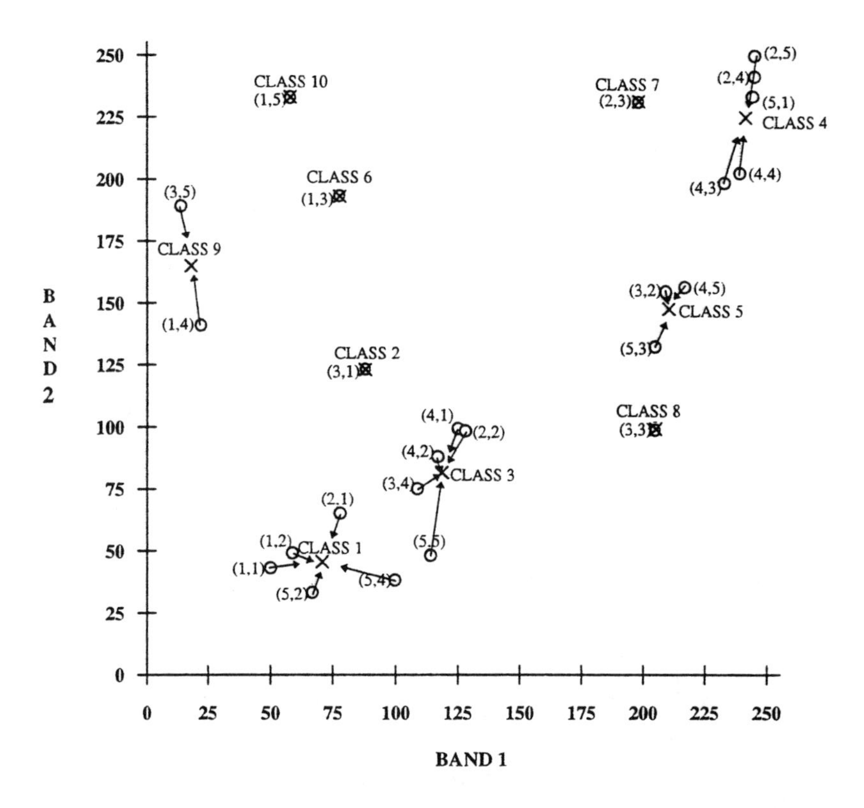

Figure 6.9. Pixel-by-pixel assignment of spectral class based on nearest spectral class mean.

#1. The ISODATA procedure then creates less variable spectral classes from the variable initial spectral class.

The ISODATA procedure is similar to sequential clustering in that the user generally specifies the clustering criteria. These criteria typically are listed in Table 6.5. One variation of the ISODATA procedure is as follows:

Pass #1

1) Check whether any classes should be split. Split any class that is too variable (the standard deviation exceeds the maximum allowed). Split the class by creating two new class centers ± one standard deviation from the mean. For example, assume we set the maximum standard deviation at 50. If class one has means of X = 100 and Y = 150 with standard deviations of SDx = 20 and SDy = 60, we would split this class into 2 spectral classes with new class centers at X = 100, Y = 90 and X = 100, Y = 210. Assign members to each class based on the closest spectral distance. Calculate new means and standard deviations.
2) Check whether spectral classes should be merged. Merge any spectral classes that are too close (distance between class means is less than minimum specified by user), or too small (number of members in a class is less than

1	1	2	3	4
1	3	5	3	1
6	7	8	4	5
9	4	3	4	1
10	4	9	5	3

Classified Image Values

Color Palette:

1 = White (W)	6 = Light Green (LG)
2 = Yellow (Y)	7 = Dark Blue (DB)
3 = Orange (O)	8 = Light Blue (LB)
4 = Red (R)	9 = Brown (Br)
5 = Dark Green (DG)	10 = Black (Bl)

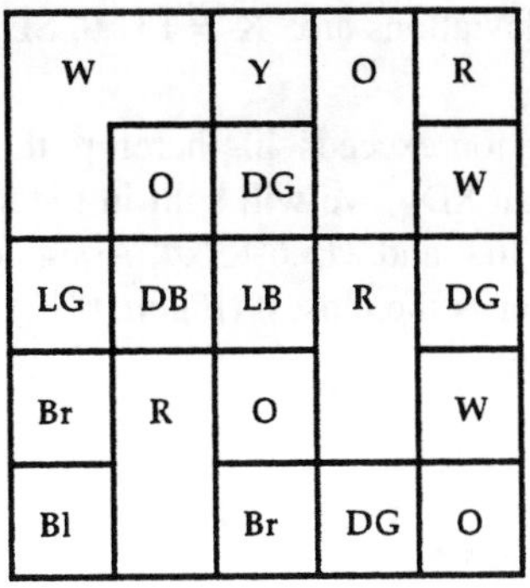

W		Y	O	R
	O	DG		W
LG	DB	LB	R	DG
Br	R	O		W
Bl		Br	DG	O

Displayed Classified Image

Figure 6.10. Classified image displayed using color assignments from color palette.

minimum specified by user). Also merge the closest classes if the total number of spectral classes exceeds the maximum specified by the user. Calculate new means and standard deviations.

3) Check whether ISODATA procedure should stop. Stop if the maximum number of iterations has been reached or the percent of spectral class members that are unchanged is greater than the user-specified value.

Table 6.5. Example of criteria set by user for ISODATA clustering

Starting Criteria
- Initial number of spectral classes

Processing Criteria
- Splitting criterion:
  - maximum variation of spectral class
- Merging criteria:
  - minimum distance between spectral class means
  - maximum number of spectral classes
  - minimum number of members in a class

Stopping Criteria
- Maximum number of iterations
- Desired percent unchanged class members

Pass #2

1) Once the mean digital values for spectral classes have been determined from pass #1, classify each pixel in the image. Start at pixel 1,1 and process the image sequentially, pixel by pixel.

Let's try an example by using the data from Figure 6.6. First we specify some criteria: initial number of spectral classes = 1, maximum allowed class standard deviation = 40, minimum distance between class means = 5, maximum number of spectral classes = 5, minimum number of members in a class = 2, maximum number of iterations = 3, desired percent unchanged class members = 90%.

Iteration #1: We have one spectral class consisting of all image pixels. The means and standard deviations are: $\overline{X}$ = 137.9, SDx = 77.7 $\overline{Y}$ = 136.4, SDy = 72.5 (Figure 6.11).

The standard deviation exceeds 40, therefore this class will be split. Since the SDx is greater than SDy, we will split in the x direction. The new class centers will be 60.2,136.4 and 215.6,136.4. Assign each pixel as a member to which ever class center is the closest (Figure 6.12). Calculate the new means and standard deviations.

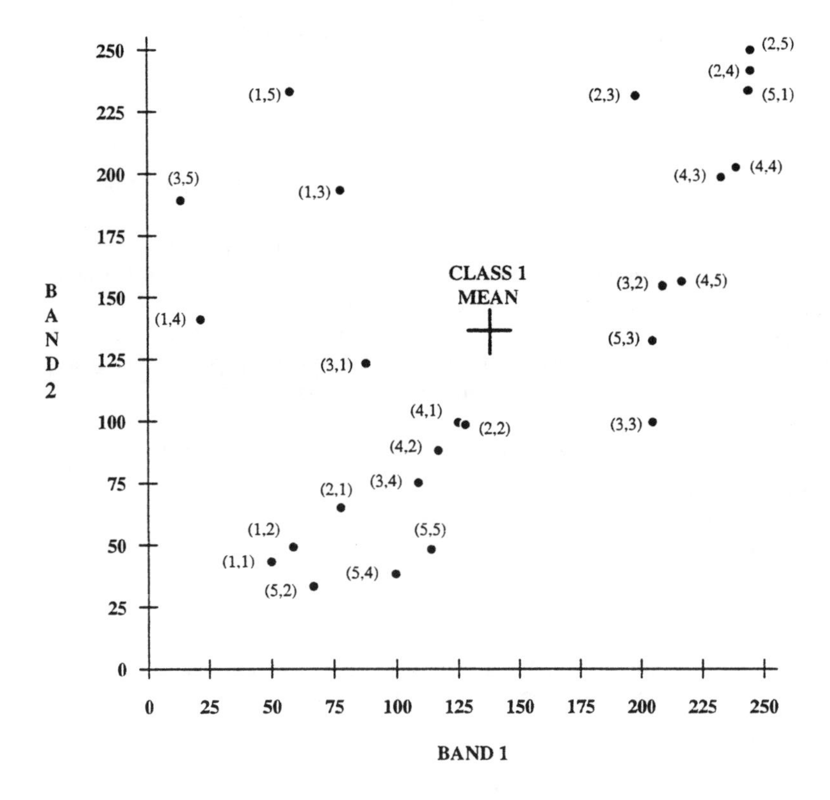

Figure 6.11. Single spectral class at the beginning of ISODATA clustering

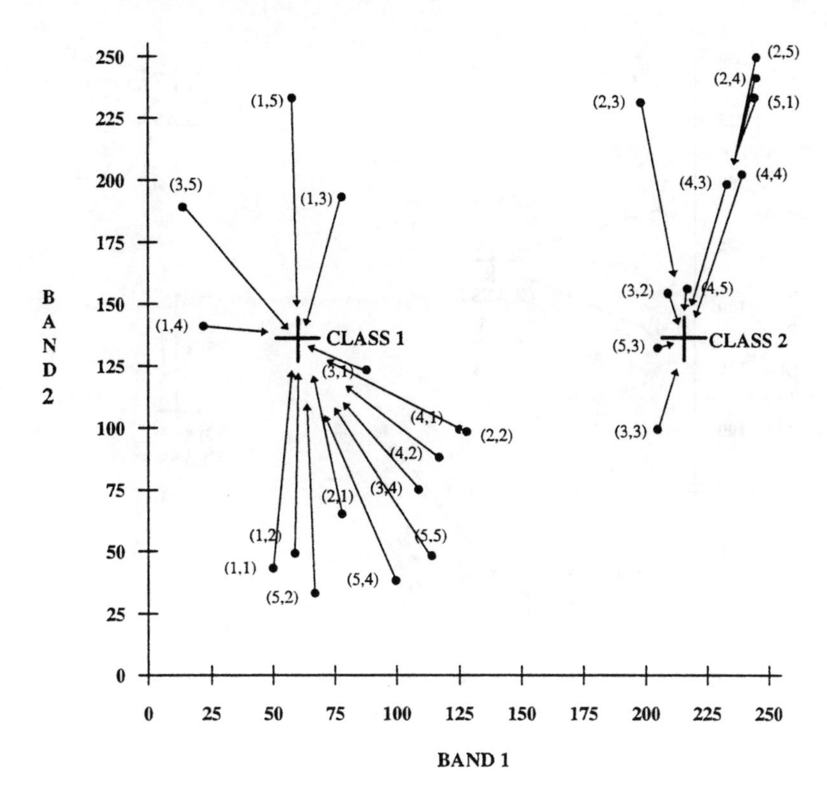

Figure 6.12. Split of single spectral class during ISODATA iteration #1.

Check if any classes should be merged: Are there spectral classes with spectral distance between means less than five? Are there more than five spectral classes? Are there spectral classes with only one member? No, so we do not merge any spectral classes.

Check stopping criteria. Is the number of iterations equal to three? Are the percent class members unchanged greater than 90%? No, so we continue with iteration #2.

Iteration #2: Check if any classes should be split. The classes have the following values (Table 6.6). Both classes have standard deviations that exceed our limit of 40. Therefore we assign new class centers of X = 81.4, Y = 43.6 X = 81.4,Y = 168.2 X = 209.7,Y = 107.3 X = 209.7,Y = 243.3. We then each pixel to the nearest class center (Figure 6.13). Then we calculate the means and standard deviations for each class.

Table 6.6. Spectral class statistics at beginning of ISODATA iteration #2

Spectral Class	Mean X	Standard Deviation X	Mean Y	Standard Deviation X
1	81.4	36.9	105.9	62.3
2	209.7	50.7	175.3	68.0

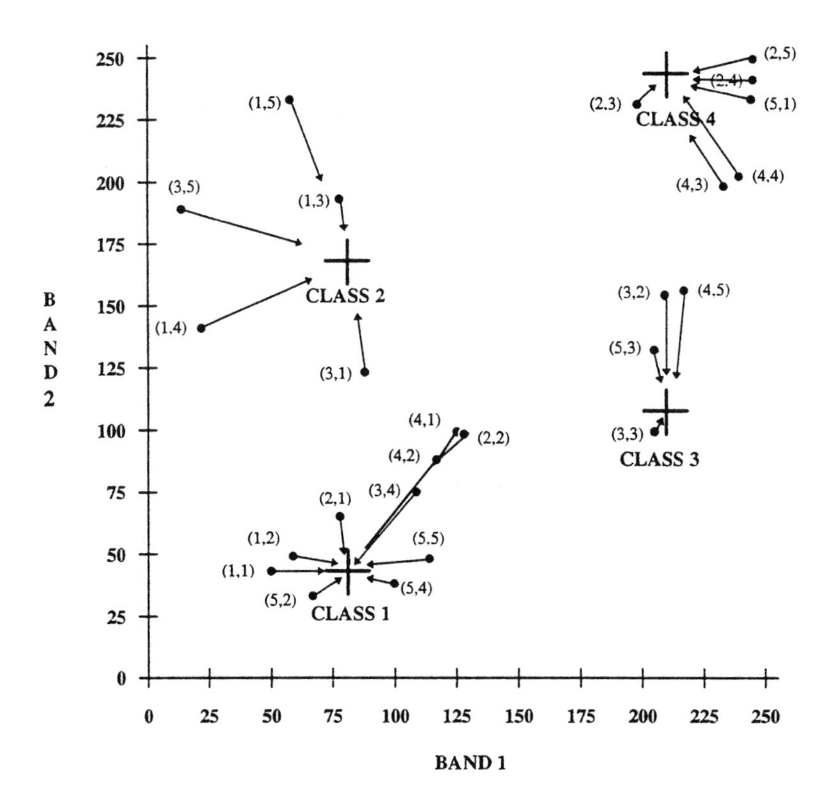

Figure 6.13. Spectral class assignments after splitting phase of ISODATA iteration #2.

Check if any classes should be merged: Are there spectral classes with spectral distance between means less than five? Are there more than five spectral classes? Are there spectral classes with only one member? No, so we do not merge any spectral classes.

Check stopping criteria. Is the number of iterations equal to three? Are the percent class members unchanged greater than 90%? No, so we continue with iteration #3.

Iteration #3: Once again, determine whether any spectral classes are too variable and therefore should be split (Table 6.7).

Class 2 is too variable in the Y-direction (SDy > 40), and therefore we assign new class centers of X = 52.0, Y = 131.8 and X = 52.0, Y = 219.8. Assign

Table 6.7. Spectral class statistics at beginning of ISODATA iteration #3

Spectral Class	Mean X	Standard Deviation X	Mean Y	Standard Deviation X
1	94.7	28.8	63.6	25.0
2	52.0	33.0	175.8	44.0
3	209.0	5.7	135.2	26.5
4	234.0	18.2	225.7	20.9

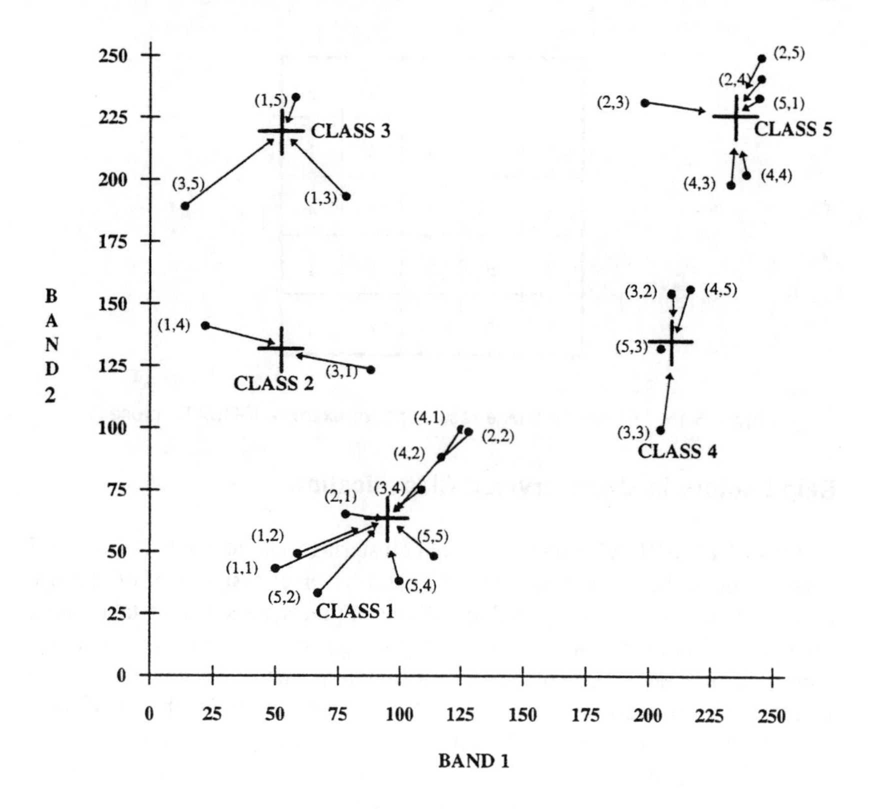

Figure 6.14. Spectral class assignments after splitting phase of ISODATA iteration #3.

class members based on minimum spectral distance (Figure 6.14) and calculate new means and standard deviations (Table 6.8).

Check if any classes should be merged: Are there spectral classes with spectral distance between means less than five? Are there more than five spectral classes? Are there spectral classes with only one member? No, so we do not merge any spectral classes.

Check stopping criteria. Is the number of iterations equal to three? Yes, so we are done with Pass #1! Using Table 6.8, we can assign every pixel to the closest spectral class means during Pass #2. The resulting classified image would have the following spectral class values (Figure 6.15.)

Table 6.8 Spectral class statistics at the end of iteration #3

Spectral Class	Mean X	Standard Deviation X	Mean Y	Standard Deviation X
1	94.7	28.8	63.6	25.0
2	55.0	46.7	132.0	12.7
3	50.0	32.7	205	24.3
4	209.0	5.7	135.2	26.5
5	234.0	5.7	135.2	26.5

1	1	2	1	5
1	1	4	1	1
3	5	4	5	4
2	5	1	5	1
3	5	3	4	1

Figure 6.15. Classified image resulting from example ISODATA process

Skip Factors in Unsupervised Classification

Pass #1 of ISODATA and sequential clustering generates a list of spectral class means to be used during pass #2 pixel-by-pixel assignment of spectral classes. If our image is composed of millions of pixels, pass #1 may take a long time. Therefore many image processing packages allow you to specify a skip factor for pass #1. The skip factor is a sampling factor; for example a skip factor of 10,10 means sample every tenth column and every tenth row. If there are tight clusters in the data, a classified image that was quickly clustered using a large skip factor can be very similar to a classified image that was clustered with every pixel being sampled during pass #1.

GROUPING OF SPECTRAL CLASSES

We may have many spectral classes resulting from unsupervised classification. Some of these spectral classes may be from the same cover type. For example, imagine that our goal is to produce a forest type map and we have 5 spectral classes that correspond to aspen. This may be due to differing densities, stand sizes, topographic positions, etc. of various aspen stands. There are several tools that we can use for examining which spectral classes should be aggregated into a cover class; these tools can use to examine spatial or spectral similarity of classes.

Grouping Based on Spatial Similarity

There are three common ways to examine a classified image for spatial similarity of spectral classes: 1) cursor inquiry, 2) color palette manipulations, and 3) overlay of spectral classes on the rectified image.

Cursor Inquiry

A cursor inquiry program can be used to examine the ground-based coordinates (e.g., UTM or state plane) and spectral class number of each pixel within a classified image. Suppose we know the map coordinates location of several aspen stands from interpretation of aerial photography. We could use a cursor inquiry program to examine these stand locations on the classified image and determine all the spectral classes that are within these stands. We could also use GPS in the field to determine the UTM coordinates of the centers of aspen stands. We could then use a cursor inquiry program to determine the spectral class numbers of the pixels close to these UTM coordinates. One disadvantage of the cursor inquiry approach is that we are biased in examining spectral classes — we typically look only at areas where we have good cover-type information. For example, by using a cursor inquiry program, we might find that all known lakes were classified into spectral classes 19 and 75. However, cloud shadows, shadowed canyons, and recently burned areas may also have been classified into these two spectral classes.

Color Palette Manipulation

It is important to look at the distribution of spectral classes across the entire classified image — not just in the areas we are familiar with. This can easily be done by changing the color of user-specified spectral classes and examining where these classes occur on the image. For example, we could examine the distribution of spectral classes 19 and 75 by changing the colors assigned to these classes; we might color these two classes a bright red and examine where they occurred within the image. We could then compare these areas with aerial photographs or maps to determine whether spectral classes correspond to known cover types.

Spectral Class Overlay

Another spatially-based technique is to display spectral classes over the image. We could then visually interpret the image to determine where certain cover types occur. For example, we could display a SPOT HRV multispectral image as a simulated color infrared photograph by assigning bands 3,2,1 to the red, green, and blue planes. Aspen stands (broadleafs) would typically appear pinkish red and spruce stands (conifers) would appear purple on such an image. We could then use the graphics plane to display selected spectral classes over the original image. For example, we could display all 5 classes we thought corresponded to aspen and examine how well they correspond to the pink aspen stands on the color infrared image.

Grouping Based on Spectral Similarity

Suppose we performed an unsupervised classification with a seven band image to produce 100 spectral classes. At first, we might want to consider grouping classes together that were *spectrally* similar to one another. Then we could consider whether spectrally similar classes correspond to the same vegetation cover type. If we had a picture portraying the spectral similarity of all 100 spectral classes, we could use the spatially based techniques to consider the grouping of similar spectral classes. Such a picture is called a dendrogram, and it is produced by a simple procedure called hierarchical cluster analysis.

Let's look at a simple example. Imagine that we perform an unsupervised classification on a two-band image. The mean digital values for the 5 spectral classes are shown in Table 6.9.

Table 6.9. Spectral classes to be grouped

	Mean Digital Values	
Spectral Class	**Band #1**	**Band #2**
1	10	5
2	20	20
3	30	55
4	30	40
5	50	90

Which of these classes are spectrally the most similar? We can use hierarchical cluster analysis to answer this question. Hierarchical cluster analysis is an iterative procedure that can be performed as follows:

Step 1) Compute a resemblance matrix which describes the spectral (or Euclidean distance) between all spectral class means. For example, the spectral distance between class 1 and class 2 is:

$$SQRT[(10 - 20)^2 + (5 - 20)^2] = 18.0$$

We could compute the spectral distance between all classes and fill in a table (called a resemblance matrix) shown in Table 6.10. Notice that the empty cell values can be mirrored from the existing values (the distance between 2 and 3 is the same as the distance between 3 and 2).

Step 2) Group the two most similar spectral clusters. In this example, spectral classes 3 and 4 are the most similar with a spectral distance of

Table 6.10. Initial resemblance matrix

Spectral Class	1	2	3	4	5
2	18.0	—			
3	53.9	36.4	—		
4	40.3	22.4	15.0*	—	
5	93.9	76.2	40.3	58.9	—

15.0. Therefore we have a new cluster consisting of classes 3 and 4 grouped into one class. Let's call this new class (34). Class (34) has a mean digital value of 30 for band1 and 12.5 for band 2.

Step 3) Since we have a new grouped spectral class, we recompute another resemblance matrix (Table 6.11).

Table 6.11. Resemblance matrix after first clustering

Spectral Class	1	2	5	(34)
2	18.0*	—		
5	93.9	76.2	—	
(34)	47.0	29.3	47.0	—

Step 4) Group the two most similar spectral classes. Classes 1 and 2 have the shortest spectral distance (18.0) and therefore will be aggregated. This new cluster will be called class (12) and has mean digital values of 15.0,12.5 for band 1 and band 2.

Step 5) Recompute the resemblance matrix for the remaining spectral classes (Table 6.12).

Table 6.12. Resemblance matrix after second clustering

Spectral Class	(12)	(34)	5
(34)	38.1*	—	
5	85.0	47.0	—

Step 6) Group the two most similar classes. In this case, spectral classes (12) and (34) are grouped (spectral distance of 38.1) and a new cluster mean is computed as 22.5,30.0 for cluster (1234).

Step 7) We now have only two spectral classes (or clusters): clusters (1234) and 5. The spectral distance between these two classes is 66.0.

Step 8) The final step is to produce a tree or dendrogram which shows the similarity of spectral classes from the hierarchical cluster analysis procedure. We can draw the tree as shown in Figure 6.16.

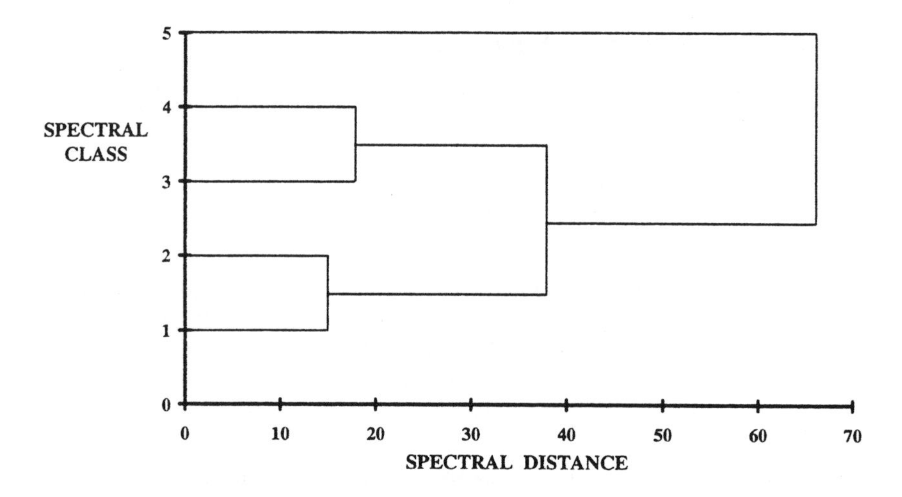

Figure 6.16. Dendrogram portraying spectral similarity among classes.

Such a dendrogram is a very useful tool when we are considering aggregating hundreds of classes from unsupervised classification. Typically we will use a dendrogram to determine which classes are spectrally similar and then apply one of the spatially based tools to examine whether classes that are spectrally similar are also spatially similar and whether these classes consistently occur in one cover type and therefore should be aggregated.

CHAPTER 6 PROBLEMS

1) You are interested in mapping pothole wetlands in North Dakota. Using the following 3-band sample image, perform a sequential clustering unsupervised classification using the following parameters:

   Maximum number of classes →4
   Maximum allowable spectral class radius →30

COLOR PALETTE	
Spectral Class 1	black
Spectral Class 2	green
Spectral Class 3	brown
Spectral Class 4	yellow

ORIGINAL DIGITAL IMAGE

5 3 1	7 4 2	210 150 123	198 173 187
6 4 0	8 4 2	55 123 189	59 129 174
55 34 29	92 19 123	39 29 23	253 198 201

COLORED CLASSIFIED IMAGE

2) Using the image from question 1, perform the same sequential clustering, this time use only bands 1 and 2.

3) Suppose you do an unsupervised classification on a SPOT multispectral image. You generate 5 spectral classes with the following mean digital values:

Spectral Class:	Mean Digital Values:		
	Band 1	Band 2	Band 3
1	34	23	4
2	69	133	198

3	67	147	231
4	139	59	182
5	167	59	209

Perform hierarchical cluster analysis and draw a dendrogram to portray the spectral similarities among these five classes.

4) Perform an ISODATA unsupervised classification using bands 1 and 2 from question#1 using the following criteria:

Starting Criteria Initial number of spectral classes: 1
Processing Criteria Splitting criterion: maximum standard deviation of spectral class: 10 Merging criteria: minimum distance between spectral class means: 25 maximum number of spectral classes: 3 minimum number of members in a class: 2
Stopping Criteria Maximum number of iterations: 3 Desired percent unchanged class members: 90

5) Read "A Critical Review and a New Approach to Unsupervised Multispectral Classification" by Ron Wazowski (*1992 ASPRS Annual Convention Technical Papers,* pp. 519–528.)

Wazowski criticizes sequential clustering unsupervised procedures as being highly unstable under certain circumstances. This is because (check the best answers):

____ different spectral classes can result if different skip factors are used for pixel sampling during pass #1.

____ different spectral classes can result if the image is processed from left to right versus from right to left during pass #1.

____ spectral classes can be different if the exactly the same parameters and procedures are used on 2 different images.

____ all spectral bands have equal weight in sequential clustering, yet some bands may be significantly more important for accurate delineation of cover classes.

ADDITIONAL READINGS

General

Bryant, J. 1979. On the clustering of multidimensional pictorial data. *Pattern Recognition.* 11:115–125.

Chuvieco, E. and R. G. Congalton. 1988. Using cluster analysis to improve the selection of training statistics in classifying remotely sensed data. *Photogrammetric Engineering and Remote Sensing.* 54:1275–1281.

Dubes, R. and A. K. Jain. 1976. Clustering techniques: The user's dilemma. *Pattern Recognition.* 8:247–260.

Goldberg, M. and S. Shlien. 1976. A four-dimensional histogram approach to the clustering of Landsat data. *Canadian Journal of Remote Sensing.* 2:1–11.

Justice, C. and J. Townshend. 1982. A comparison of unsupervised classification procedures on Landsat MSS data for an area of complex surface conditions in Basilicata, southern Italy. *Remote Sensing of Environment.* 12:407–420.

McGwire, K. C. 1992. Analyst variability in labeling of unsupervised classifications. *Photogrammetric Engineering and Remote Sensing.* 1673–1677.

Robinove, C. J. 1981. The logic of multispectral classification and mapping of land. *Remote Sensing of Environment.* 11:231–244.

Wharton, S. W. and B. J. Turner. 1981. ICAP: An interactive cluster analysis procedure for analyzing remotely sensed data. *Remote Sensing of Environment.* 11:279–293.

Applications

Franklin, J. 1993. Discrimination of tropical vegetation types using SPOT multispectral data. *Geocarto International.* 8:57–63.

Homer, C. G., Edwards, T. C., and R. D. Ramsey. 1993. Use of remote sensing methods in modelling sage grouse winter habitat. *Journal of Wildlife Management.* 57: 78–94.

Loveland, T. R., Merchant, J. W., Ohlen, D. O. and J. F. Brown. 1991. Development of a land-cover characteristics database for the conterminous U.S. *Photogrammetric Engineering and Remote Sensing.* 57:1453–1463.

Niemann, K. O. 1991. Landscape drainage modelling to enhance landsat classification accuracies. *Geocarto International.* 6:13–30.

Plumb, G. A. 1991. Assessing vegetation types of Big Bend National Park, Texas for image-based mapping. *Vegetatio.* 94:115–124.

Price, K. P., Pyke, D. A., and L. Mendes. 1992. Shrub dieback in a semiarid ecosystem: the integration of remote sensing and geographic information systems for detecting vegetation change. *Photogrammetric Engineering and Remote Sensing.* 58: 455–463.

Rutchey, K. and L. Vilcheck. 1994. Development of an everglades vegetation map using a SPOT image and the global positioning system. *Photogrammetric Engineering and Remote Sensing.* 60:767–775.

Stenback, J. and R. Congalton. 1990. Using Thematic Mapper imagery to examine forest understory. *Photogrammetric Engineering and Remote Sensing.* 56:1285–1290.

CHAPTER 7

Supervised Classification

INTRODUCTION

The basic strategy in supervised classification is to sample areas of known cover types to determine representative spectral values of each cover type. These sample areas are generally referred to as *training fields* and the representative spectral values from these training fields are sometimes called *spectral signatures*. Once representative spectral values have been established for each cover type, an image can then be classified. Each pixel is predicted or classified based on its similarity to representative cover type spectral values.

Supervised classification has several potential advantages over unsupervised classification. In unsupervised classification, the user selects fairly subjective parameters such as number of spectral classes, skip factor, and clustering criteria. The user must then decide what cover type corresponds to each spectral class. In supervised classification, the user determines the number of cover classes and what those cover classes are *before* classification. A potential problem with unsupervised classification is that some spectral classes may be composed of several cover classes. For example, a spectral class from unsupervised classification might be composed of water, north-facing spruce-fir stands, and areas burned by fire. However, these cover types might be classified more accurately in supervised classification if separate training areas are selected from water, steep spruce-fir stands, and burned areas.

There are several potential disadvantages with supervised classification when compared to unsupervised classification. First, the user must have training fields for every cover type on the image. For example, if we had training fields for red pine and jack pine stands, but not white pine stands, all white pine stands would be misclassified.

Second, training fields are selected based on their cover types and not necessarily their spectral differences. For example, we might select training fields for jack pine from only plantations of medium density. Therefore other jack pine stands with lower density might be missclassified as low density white pine, while high density jack pine stands might be missclassified as high density spruce stands. This type of problem is avoided in unsupervised classification since spectrally similar pixels are automatically grouped into the same spectral class.

TRAINING FIELDS

A training field is a sample area for estimating representative spectral statistics of a certain cover type. For example, if we were interested in forest type mapping in the north central United States, we might establish training fields in jack pine, red pine, spruce-fir, northern hardwoods, aspen/birch, and oak stands. You can think of training fields as representative samples. Since we rely on these training fields to estimate typical spectral values for each cover type, it is extremely important to establish training fields from homogeneous cover type areas.

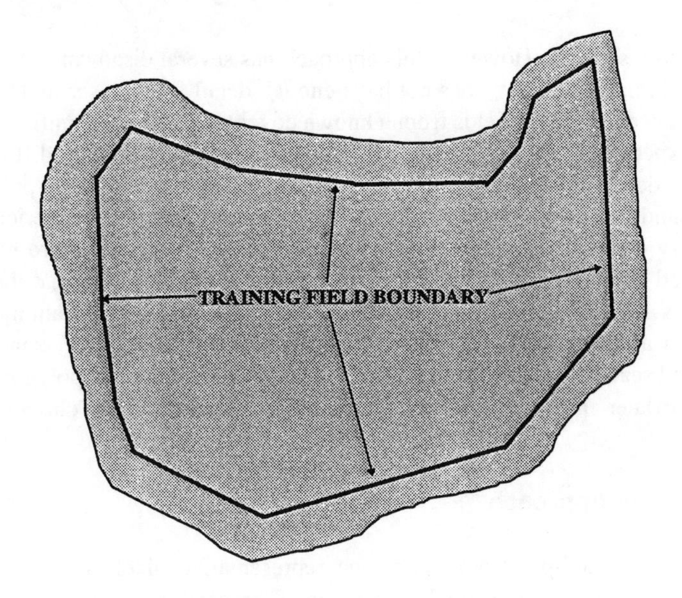

Figure 7.1. Map digitizing of a jack pine training field.

There are several approaches that can be used in establishing training fields including map digitizing, on-screen digitizing, and a seed-pixel approach.

Map Digitizing

Training fields are commonly transferred from aerial photographs to base maps and then digitized from maps using a digitizing tablet. The advantage of this approach is that each training field can be closely examined by stereoscopic viewing of aerial photographs. Also, the photos and maps can easily be checked in the field. A disadvantage of the approach is that since the digital image has positional error (image rectification is not perfect), there will be some misregistration between the location of the training fields boundaries and the same locations on the digital image. This positional error may be due to errors in transferring training fields to the base map, error in registering the base map to a coordinate system, error in digitizing training field polygons, and positional errors inherent in the digital image. Because of these potential problems, it is wise to delineate training fields from the interior of large homogeneous stands and to avoid "edge pixels" (Figure 7.1).

On-Screen Digitizing

Another potential problem with the map digitizing approach is that the training field cover types may differ between old maps or aerial photographs and recent satellite imagery. This disadvantage can be avoided by on-screen digitizing of training fields directly from the digital image. Training fields polygon boundaries can be simply traced on the screen with most image processing systems. However, this approach has several disadvantages. First, a digital image typically does not have enough detail for the user to delineate homogeneous training fields from a known cover type without significant user field experience in the study area. For example, it would be very difficult to visually delineate red pine versus jack pine versus white pine training polygons on a Landsat Thematic Mapper image without significant field experience or ancillary data such as large scale aerial photography. Second, there may be scattered pixels of a different cover type within a training field that the user assumes is homogeneous. For example, we might digitize an training field polygon assumed to be pure aspen/birch and yet we may have erroneously included scattered pixels composed of sedge meadow within the polygon. You will see later in this chapter that this could drastically affect classification results.

Seed-Pixel Approach

With the seed-pixel approach, one representative pixel is chosen as a starting pixel for training field delineation. Then candidate pixels around the seed pixel are sequentially considered as possible additional training-field pixels. Those pixels that are spectrally similar to the seed pixel are included as training-field pixels; while pixels that are not spectrally similar are rejected as training field pixels. The major advantage of the seed-pixel approach is that it helps to delineate spectrally homogeneous training fields. For example, imagine that the following is a potential training field consisting mainly of aspen/birch (AB) but also interspersed with sedge meadow (SM) pixels (Figure 7.2).

If we knew that pixel 6,3 was definitely from the center of an aspen/birch stand, we would use that pixel as a seed pixel. We next would have to define a critical spectral distance for surrounding candidate pixels. How close is "close enough" for a pixel to be included as an additional training-field pixel? This spectral distance is specified by the user. For example, if we specified the critical spectral distance as 5, any surrounding contiguous pixels that had a spectral distance less than or equal to 5 would be included as training-field pixels. The training field would "grow" from the seed pixel until all surrounding candidate pixels beyond the user-specified spectral distance of 5. When the seed-pixel procedure is completed the training field would look like Figure 7.3.

The critical spectral distance is usually unknown to the user. Typically, the user tests a variety of spectral distances before deciding the best spectral distance for defining a training field using the seed-pixel approach. For example, if we used a critical spectral distance of 1 we would find that very few aspen/birch pixels were selected as training-field pixels. On the other hand, if we used a critical spectral distance of 100, we would find that the sedge/meadow pixels would be erroneously included as training-field pixels.

Once training fields have been established for each cover type, they can be used to generate spectral statistics for classifying the entire image. There are many different classifiers used in supervised classification; in this chapter you will learn about several commonly used classifiers.

POPULAR SIMPLE CLASSIFIERS

Minimum Distance Classifier

We could easily compute the mean digital values for each training field and use these means to predict the cover class of each image pixel. Imagine that we

151 154 (AB)	152 157 (AB)	153 155 (AB)	153 154 (AB)	155 159 (AB)	180 160 (SM)	152 156 (AB)	183 195 (SM)	185 199 (SM)	205 214 (SM)
152 155 (AB)	154 157 (AB)	200 210 (SM)	154 158 (AB)	151 154 (AB)	152 155 (AB)	150 152 (AB)	152 155 (AB)	151 154 (AB)	203 209 (SM)
154 157 (AB)	151 153 (AB)	200 209 (SM)	206 213 (SM)	153 157 (AB)	**153** **156** **(AB)**	152 156 (AB)	199 202 (SM)	152 156 (AB)	153 156 (AB)
155 158 (AB)	199 203 (SM)	153 155 (AB)	150 152 (AB)	152 156 (AB)	152 155 (AB)	153 154 (AB)	201 211 (SM)	151 155 (AB)	156 160 (AB)
154 157 (AB)	201 207 (SM)	153 156 (AB)	240 245 (SM)	245 252 (SM)	155 159 (AB)	200 240 (SM)	205 235 (SM)	152 155 (AB)	151 155 (AB)

SEED PIXEL

Figure 7.2. Example digital image for delineation of training-field pixels using the seed-pixel approach.

<table>
<tr><td>151
154
(AB)</td><td>152
157
(AB)</td><td>153
155
(AB)</td><td>153
154
(AB)</td><td>155
159
(AB)</td><td>180
160
(SM)</td><td>152
156
(AB)</td><td>183
195
(SM)</td><td>185
199
(SM)</td><td>205
214
(SM)</td></tr>
<tr><td>152
155
(AB)</td><td>154
157
(AB)</td><td>200
210
(SM)</td><td>154
158
(AB)</td><td>151
154
(AB)</td><td>152
155
(AB)</td><td>150
152
(AB)</td><td>152
155
(AB)</td><td>151
154
(AB)</td><td>203
209
(SM)</td></tr>
<tr><td>154
157
(AB)</td><td>151
153
(AB)</td><td>200
209
(SM)</td><td>206
213
(SM)</td><td>153
157
(AB)</td><td>153
156
(AB)</td><td>152
156
(AB)</td><td>199
202
(SM)</td><td>152
156
(AB)</td><td>153
156
(AB)</td></tr>
<tr><td>155
158
(AB)</td><td>199
203
(SM)</td><td>153
155
(AB)</td><td>150
152
(AB)</td><td>152
156
(AB)</td><td>152
155
(AB)</td><td>153
154
(AB)</td><td>201
211
(SM)</td><td>151
155
(AB)</td><td>156
160
(AB)</td></tr>
<tr><td>154
157
(AB)</td><td>201
207
(SM)</td><td>153
156
(AB)</td><td>240
245
(SM)</td><td>245
252
(SM)</td><td>155
159
(AB)</td><td>200
240
(SM)</td><td>205
235
(SM)</td><td>152
155
(AB)</td><td>151
155
(AB)</td></tr>
</table>

Figure 7.3. Example digital image for delineation of training field pixels using the seed-pixel approach (user-specified spectral distance = 5).

have a 2-band image from a small area that is dominated by aspen/birch stands and sedge meadows. From training fields we compute mean digital values as shown in Table 7.1.

Using the minimum distance to means classifier, we could predict the cover type of each pixel based on its spectral distance aspen/birch versus sedge meadow means. For example, imagine that the first pixel in the image had digital values of 180,180. We could determine the minimum distance to cover class means using a ruler with Figure 7.4.

Typically, more than two spectral bands are used in minimum distance classification and the distance is computed (as we computed Euclidean distance in Chapter 6) rather than measured with a ruler. The main advantage of

Table 7.1. Hypothetical spectral statistics for aspen/birch and sedge meadow training fields

	Band 1			Band 2		
	Mean	**Min/ Max**	**Standard Deviation**	**Mean**	**Min/ Max**	**Standard Deviation**
Aspen/ Birch (n = 34)	152.6	151/156	1.5	155.7	152/160	1.9
Sedge Meadow (n = 16)	203.2	180/245	17.3	212.8	160/252	22.2

the minimum distance classifier is that it is simple and fast. However, there is a major disadvantage with this classifier: we are only using spectral means and not using information about the spectral variability of the cover classes. For example, consider Figure 7.5. You can see that the sedge meadow pixels are much more variable than the aspen/birch pixels. Perhaps this is due to some sedge meadows having bright dry soil and scattered, short, dormant sedges;

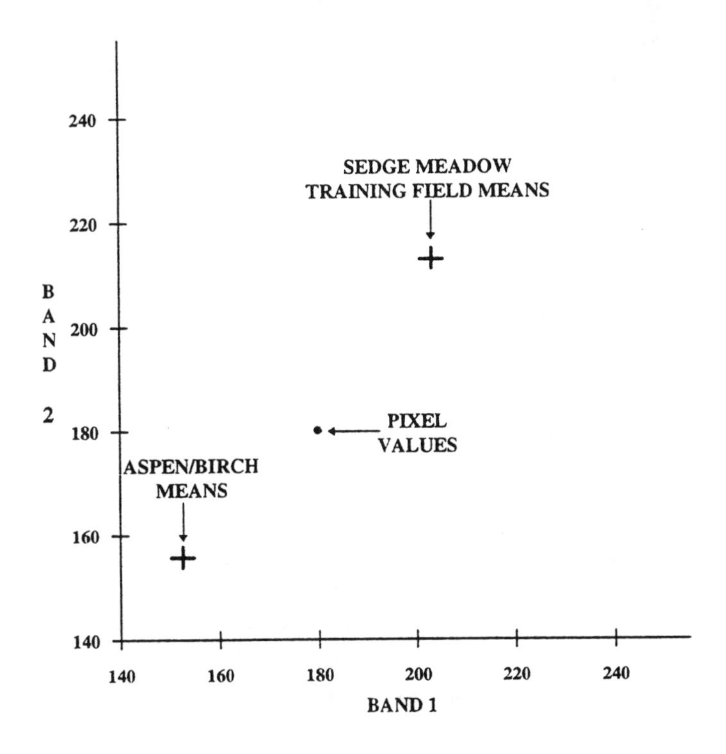

Figure 7.4 Spectral distance from candidate pixel to be classified to training field means. In this example, since the candidate pixel is closest to the aspen/birch means, it is classified as aspen/birch.

while other meadows were wet and contained tall, growing sedges. If we relied only on spectral means, we would incorrectly classify the candidate pixel as aspen/birch. However, if we included information about the variability of each cover class, we could correctly predict the candidate pixel to belong to the sedge meadow class.

Parallelepiped Classifier

We can use training fields to estimate the variability of spectral values from each cover type. This spectral variability information can then be used in a classification rule called the parallelepiped classifier. A parallelepiped is a multi-dimensional rectangle. We can build a parallelepiped for each cover class by using the cover class minimum and maximum spectral values. For example, from Table 7.1 we can use the minimum and maximum spectral values from each cover class training field to establish parallelepipeds for

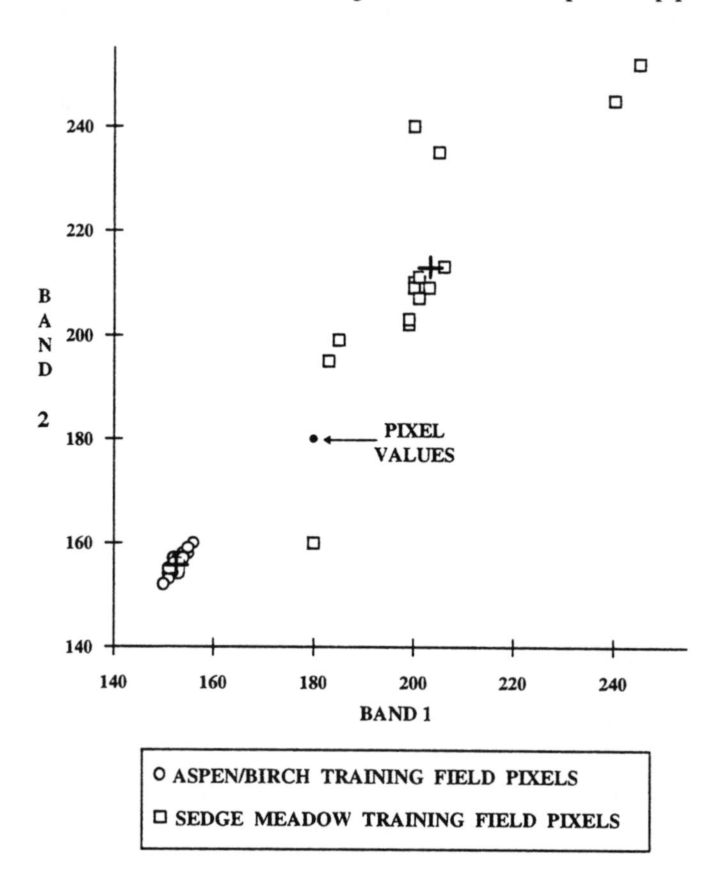

Figure 7.5. Plot showing variability of spectral values for each training field.

aspen/birch and sedge meadow (Figure 7.6). The minimum X,Y values for aspen/birch are 151,152; the maximum X,Y values are 156,160. Therefore, the lower left corner of the aspen/birch parallelepiped is at X,Y of 151,152 and the upper right corner is at X,Y of 156,160. What if a candidate pixel does not plot into any parallelepiped? Some image processing programs classify such a pixel as "unclassified," while other programs use a backup rule such as minimum distance to means for classifying pixels outside of all parallelepipeds.

One drawback with using minimum and maximum spectral values for delineating parallelepipeds is that these values may be *extremes that occur only rarely* within a given cover class (Figure 7.6). Therefore, parallelepipeds are often established using standard deviations instead of minimums and maximums. If a cover class has normally distributed spectral values, then 95% of the pixels within that cover class are expected to have spectral values within

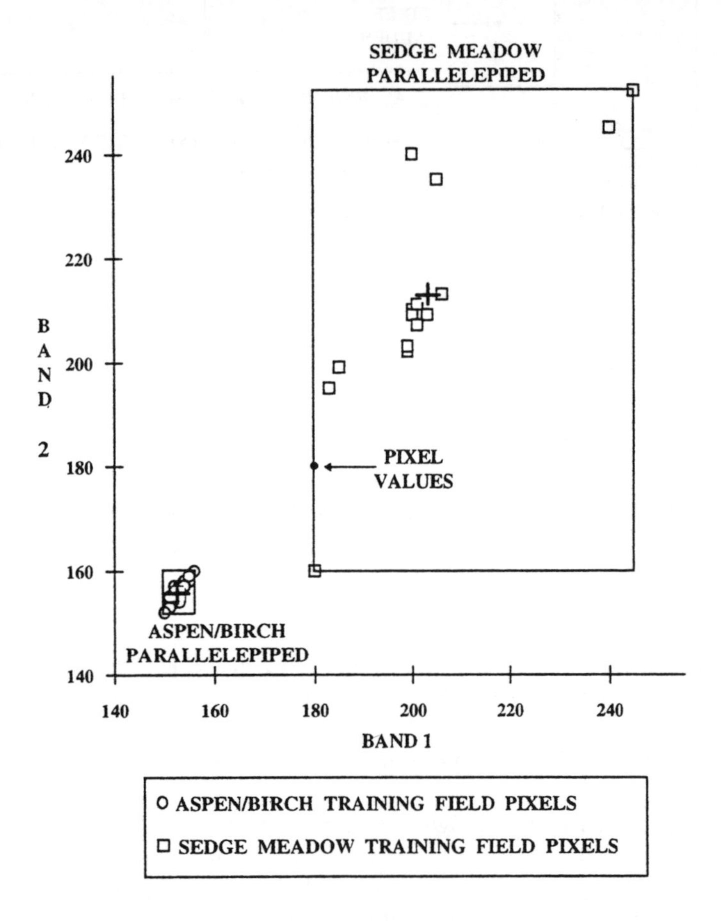

Figure 7.6. Aspen/birch and sedge meadow parallelepipeds based on minimum and maximum spectral values from training fields.

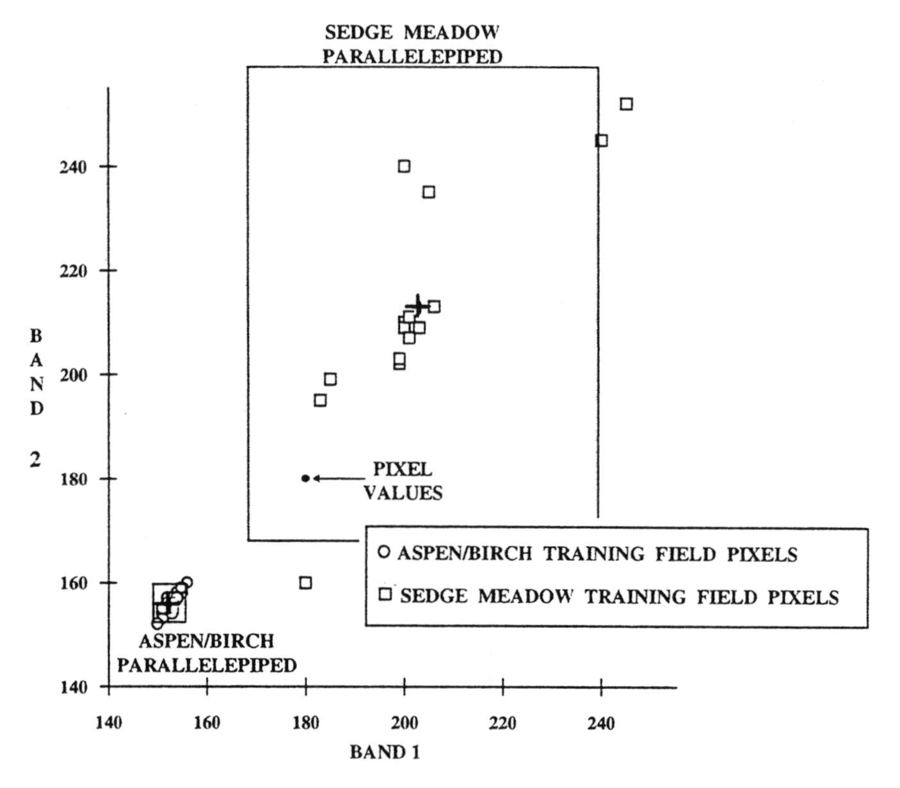

Figure 7.7. Parallelepipeds established with limits based on two standard deviations from the spectral mean.

a range of the spectral mean ±2 * standard deviation. Figure 7.7 illustrates how parallelepipeds can exclude extreme values if they are established with limits based on two standard deviations from the spectral mean. Using the statistics from Table 7.1, the corners of the parallelepipeds for aspen/birch were calculated as: lower left X,Y = 149.6,151.9 and upper right X,Y = 155.6,159.5; while the sedge meadow corners are: lower left X,Y = 168.6,168.4 and upper right X,Y = 237.8,257.2.

One potential problem with the parallelepiped classifier is that there may be overlap between two classes (Figure 7.8) due to high covariance (correlation).

Maximum Likelihood Classifier

A better classifier under high covariance conditions is the *maximum likelihood classifier*. This classifier accounts for the mean and covariance of each class by estimating the likelihood of a class at any digital value. This is analogous to generating contours of equal likelihood for each class and then comparing each candidate pixel with the likelihood to each class (Figure 7.9).

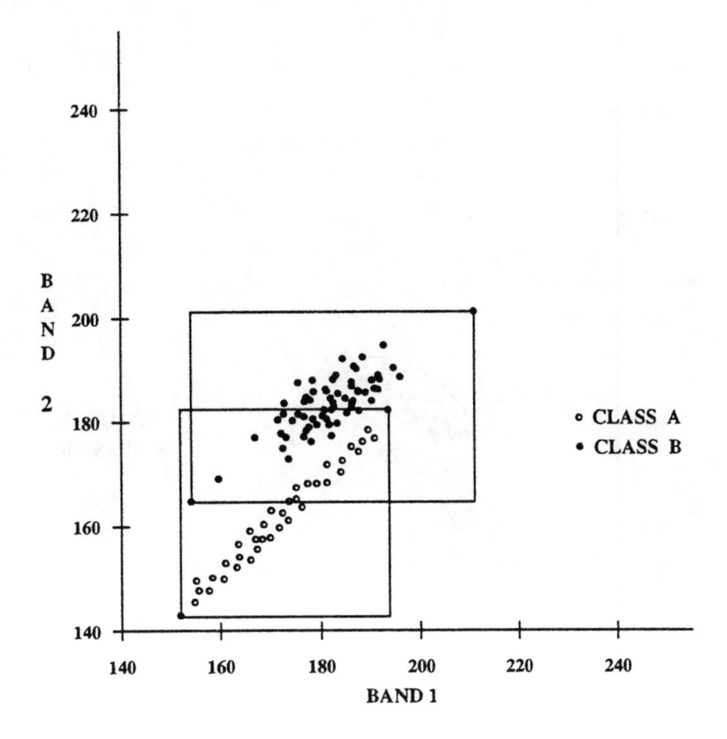

Figure 7.8. Example of overlapping parallelepipeds due to high class covariance.

The candidate pixel (band 1 = 180, band 2 = 180) is then assigned to the class with the maximum likelihood value. For example, in Figure 7.9 the highest likelihood for the candidate pixel is the likelihood that it belongs to class A. Also, notice that the likelihood peaks at each class means location.

Let's look at a simple, single-band example. Imagine that we are interested in a waterfowl management unit where two cover types dominate: cattail (CT) marsh and smartweed (SW) moist soil areas. We establish training fields for these cover types and compute the spectral statistics shown in Table 7.2.

If we assumed the distribution of digital values within each cover type to be bell-shaped or normally distributed, we can calculate the likelihood of spectral values from any given cover type. The normal distribution is given by the likelihood function:

Table 7.2. Hypothetical spectral values from cattail and smartweed training fields

	Mean Digital Value	Standard Deviation	Number of Pixels
Cattail (CT)	30	5	100
Smartweed (SW)	20	5	100

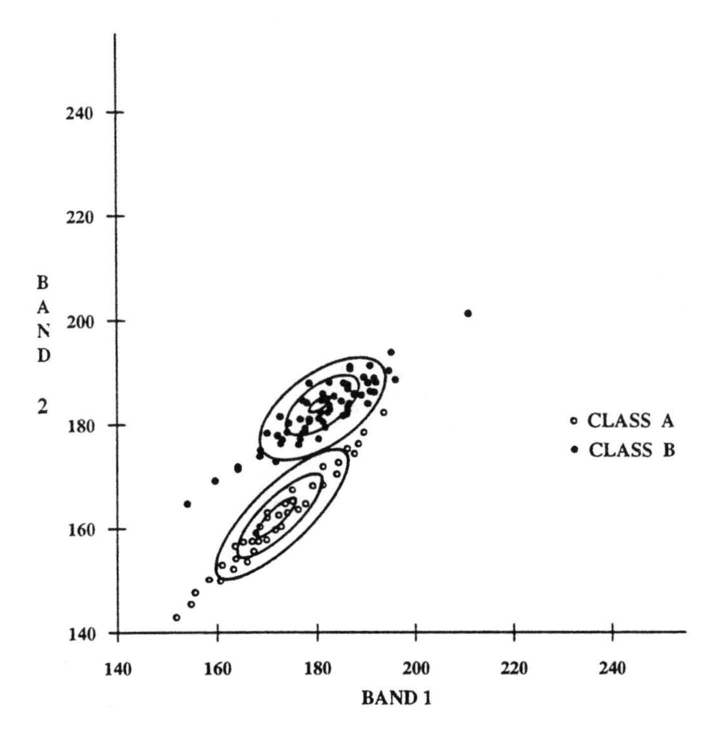

Figure 7.9. Example of contours of equal likelihood. The maximum likelihood for the candidate pixel is to belong to class A.

$$l(sv) = [\ 1/(sd * sqrt(2 * \pi))]\ exp - [\ (sv - mean)^2 / (2 * sd^2)\ ]$$

where sv is the spectral value or digital number for a given cover class, mean and sd are the mean and the standard deviation calculated from the training sample. The exp term is the inverse of the natural log. Also notice that the term $-[(sv - mean)^2/(2 * sd^2)]$ always results in a negative value. A mistake students commonly make is to forget the negative sign for this expression.

We can compute the likelihood values for cattail spectral values by using this formula. For example,

$$\begin{aligned} pr(15) &= [1/(sd * sqrt(2 * \pi))]\ exp - [(sv - mean)^2 / (2 * sd^2)] \\ &= [1/(5 * sqrt(2 * 3.14159))]\ exp - [(15 - 30)^2 / (2 * 5^2)] \\ &= [0.07978]\ exp - [4.5] \\ &= 0.0009 \end{aligned}$$

Using the same formula we can compute the likelihood values for cattail as shown in Table 7.3.

Table 7.3. Cattail likelihood values computed for a range of digital values

Spectral Value	Likelihood
10	0.00003
15	0.0009
20	0.011
22	0.022
24	0.039
26	0.058
28	0.074
30	0.080
32	0.074
34	0.058
36	0.039
38	0.022
40	0.011
45	0.009
50	0.00003

We can use these computed likelihood values to generate the following likelihood curve for cattail (Figure 7.10).

The area under the curve is 1. This is because there is a 100% likelihood of a cattail pixel having a value ranging from negative infinity to positive infinity. Notice that most of the cattail pixels are expected to have a digital value of 30

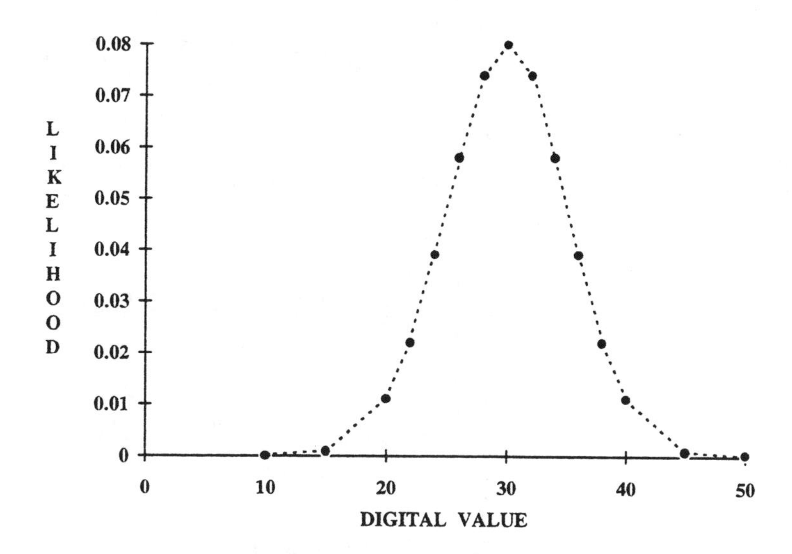

Figure 7.10. Cattail likelihood curve.

Table 7.4. **Proportions of the normal curve beyond a given standard normal deviate. For example, the proportion of a normal distribution beyond a deviate of 0 is .50 (50%), while the proportion of a normal distribution beyond a deviate of 1.00 is 0.159 (16%)**

	0.00	0.01	0.02	0.03	0.04	0.05	0.06	0.07	0.08	0.09
0.0	***0.500***	0.496	0.492	0.488	0.484	0.480	0.476	0.472	0.468	0.464
0.1	0.460	0.456	0.452	0.448	0.444	0.440	0.436	0.432	0.429	0.425
0.2	0.421	0.417	0.413	0.409	0.405	0.401	0.397	0.394	0.390	0.386
0.3	0.382	0.373	0.374	0.371	0.367	0.363	0.360	0.356	0.352	0.348
0.4	0.345	0.345	0.341	0.337	0.334	0.330	0.326	0.323	0.319	0.312
0.5	0.308	0.305	0.302	0.298	0.295	0.291	0.288	0.284	0.281	0.278
0.6	0.274	0.271	0.268	0.264	0.261	0.258	0.255	0.251	0.248	0.245
0.7	0.242	0.239	0.236	0.233	0.230	0.227	0.224	0.221	0.218	0.215
0.8	0.212	0.209	0.206	0.203	0.200	0.198	0.195	0.192	0.189	0.187
0.9	0.184	0.181	0.179	0.176	0.174	0.171	0.168	0.166	0.164	0.161
1.0	***0.159***	0.156	0.154	0.152	0.149	0.147	0.145	0.142	0.140	0.138
1.1	0.136	0.134	0.131	0.129	0.127	0.125	0.123	0.121	0.119	0.117
1.2	0.115	0.113	0.111	0.109	0.108	0.106	0.104	0.102	0.100	0.098

1.3	0.097	0.095	0.093	0.092	0.090	0.088	0.087	0.085	0.084	0.082
1.4	0.081	0.079	0.078	0.076	0.075	0.074	0.072	0.071	0.69	0.68
1.5	0.067	0.065	0.064	0.063	0.062	0.061	0.059	0.058	0.057	0.056
1.6	0.055	0.054	0.053	0.052	0.051	0.050	0.048	0.047	0.046	0.045
1.7	0.045	0.044	0.043	0.042	0.041	0.040	0.039	0.038	0.037	0.037
1.8	0.036	0.035	0.034	0.034	0.033	0.032	0.031	0.031	0.030	0.029
1.9	0.029	0.028	0.027	0.027	0.026	0.026	0.025	0.024	0.024	0.023
2.0	0.023	0.022	0.022	0.021	0.021	0.020	0.020	0.019	0.019	0.018
2.1	0.018	0.017	0.017	0.017	0.016	0.016	0.015	0.015	0.015	0.104
2.2	0.014	0.014	0.013	0.013	0.012	0.012	0.011	0.011	0.011	0.011
2.3	0.011	0.010	0.010	0.010	0.010	0.009	0.009	0.009	0.009	0.008
2.4	0.008	0.008	0.008	0.007	0.007	0.007	0.007	0.007	0.007	0.006
2.5	0.006	0.006	0.006	0.006	0.006	0.005	0.005	0.005	0.005	0.005
2.6	0.005	0.004	0.004	0.004	0.004	0.004	0.004	0.004	0.004	0.004
2.7	0.004	0.003	0.003	0.003	0.003	0.003	0.003	0.003	0.003	0.003
2.8	0.003	0.002	0.002	0.002	0.002	0.002	0.002	0.002	0.002	0.002
2.9	0.002	0.002	0.002	0.002	0.002	0.002	0.002	0.002	0.001	0.001
3.0	0.001	0.001	0.001	0.001	0.001	0.001	0.001	0.001	0.001	0.001

and pixels with values deviating from the mean are expected to be relatively rare. What percent of cattail pixels are expected to have digital values greater than 30? Half the area under the curve, or 50% of the pixels. What percent of cattail pixels are expected to have digital values greater than 35? We could attempt to solve this by using integral calculus. Fortunately, there is an easier solution; we can use a basic statistical table of deviations (Table 7.4) called *normal standard deviates* to answer this question. The normal standard deviate is the distance away from the mean expressed in units of standard deviations and can be computed as follows:

$$\frac{\text{pixel value} - \text{mean digital value}}{\text{standard deviation}}$$

Therefore, the standard normal deviate for a pixel value of 35 would be:

$$\frac{35 - 30}{5} = 1.0$$

Therefore if the cattail pixels values were normally distributed with a mean value of 30 and a standard deviation of 5, we would expect approximately 16% of the cattail pixels to have digital values greater than 35.

We can develop a likelihood curve from our smartweed training field data (assuming that the distribution of smartweed pixel values is bell shaped). The likelihood curves for both smartweed and cattail can then be used for supervised classification (Figure 7.11).

We can now classify an image based using the likelihood curves in Figure 7.11. For example, suppose we have the image shown in Figure 7.12.

The classified image (Figure 7.13) is produced by using the likelihood functions and comparing the smartweed and cattail likelihood values for each pixel. Each candidate pixel is then assigned the cover class that had the highest likelihood value. For example, a pixel with a value of 20 would have a smartweed likelihood value of 0.080 and a cattail likelihood value of 0.011; therefore the pixel would be classified as smartweed. Because we are using a single band in this example, the rule can also be simplified by defining a critical decision boundary of 25. All pixels with values less than 25 are predicted to be smartweed, and all pixels with values greater than 25 are predicted to be cattail. In this simple example, the likelihood value for smartweed and cattail is equal at a pixel value of 25. In real-life applications, this value is rarely a whole number such as 25.00000 and because digital values are integer (e.g., 24, 25, 26...) such a tie in likelihood values rarely occurs.

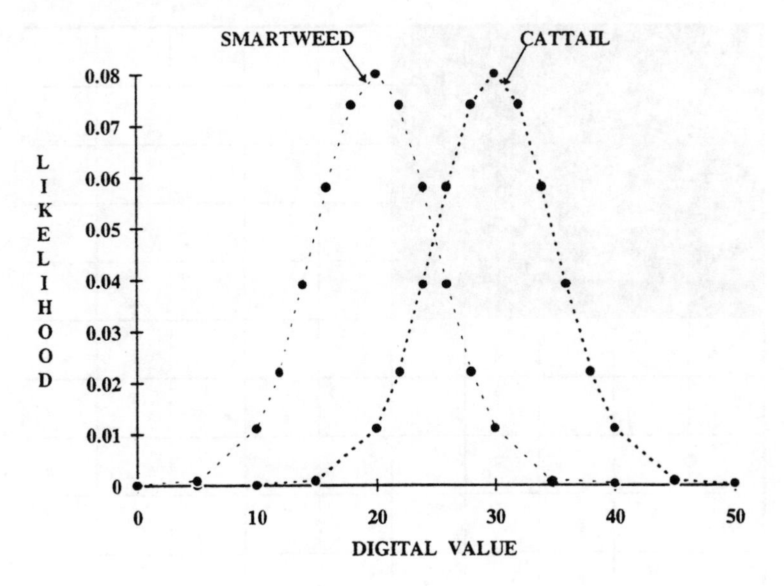

Figure 7.11. Cattail and smartweed likelihood curves developed using the formula for a normal distribution.

11	11	12	20	20	31	35	37	34	34
11	11	12	19	20	33	34	35	34	36
11	12	12	18	20	36	38	37	36	36
13	14	17	22	23	38	39	37	36	35
15	16	17	23	23	36	37	35	35	34
29	32	29	30	38	39	37	36	34	34
31	35	37	34	36	37	35	35	36	36
33	34	35	34	38	39	37	36	36	36
36	38	37	36	35	34	38	39	35	35
38	39	37	36	35	37	38	38	38	34

Figure 7.12. Hypothetical single-band image to be classified using maximum likelihood rule.

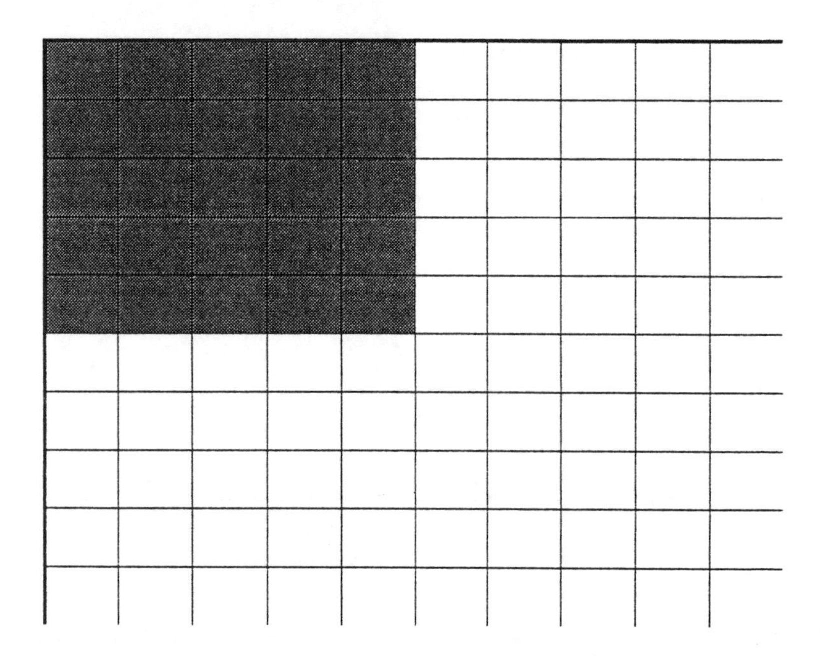

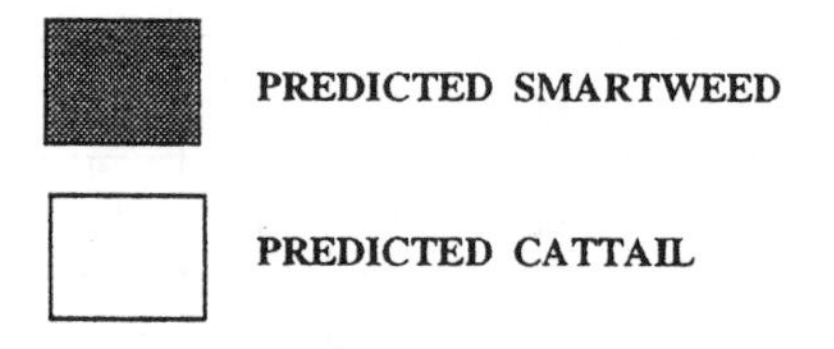

Figure 7.13. Hypothetical image classified using maximum likelihood classification rule.

CHAPTER 7 PROBLEMS

1) A landscape ecologist is trying to map old burns versus surrounding unburned tundra. The training field statistics are as follows:

Class	Band 1			Band 2		
	Mean	Min.	Max.	Mean	Min.	Max.
Burned	60.23	50	70	61.34	50	70
Unburned	100.57	60	120	90.03	65	120

Using the training field statistics, classify the following image to burned/unburned classes using the minimum distance to means classifier. Shade in the predicted burn pixels.

Original Image

80	100	70	50	55
70	70	70	60	55
80	70	100	80	60
90	90	90	70	60
55	60	65	110	110
55	50	60	90	100
55	60	60	100	120
60	60	80	110	110

Classified Image

2) Use the min/max parallelpiped classifier to classify the same image as problem #1. Shade in the predicted burn pixels.

Original Image

80	100	70	50	55
70	70	70	60	55
80	70	100	80	60
90	90	90	70	60
55	60	65	110	110
55	50	60	90	100
55	60	60	100	120
60	60	80	110	110

Classified Image

3) A moose biologist is interested in mapping moose winter range. She uses training fields of the following vegetation types:

Vegetation Type	Mean Band#1	Mean Band#2
Aspen	100.75	80.23
Black Spruce	60.39	80.28
Balsam Poplar	110.30	100.39
Willow	71.98	63.02

Using graph paper and a ruler, classify the following digital image using the minimum distance to means classifier. Shade the pixels that you classify using the following color palette assignments:

Vegetation Class	Color
Aspen	Green
Black Spruce	Black
Balsam Poplar	Blue
Willow	Red

Original Image

75,70	62,85	100,72	80,78
100,95	110,90	65,75	61,82
100,105	60,60	75,60	65,70
70,62	70,76	60,73	65,82

Classified Image

4) A waterfowl biologist wants a map of a marsh area where two vegetation types (cattail and smartweed) dominate. The mean digital values from representative training samples are:

Vegetation Type	Band#1 Mean	Band#2 Mean	Band#3 Mean	Band#4 Mean
Cattail	55	104	68	123
Smartweed	44	75	58	102

Classify the image using the minimum distance to means classifier.

SHOW YOUR CALCULATIONS! Color you classified image using the following color palette assignment:

Cattail → Yellow
Smartweed → Green

Original Image

50 81 59 111	52 77 63 120
49 100 59 120	50 98 61 107
47 79 65 108	49 99 63 120

Classified Image

5) A forester is interested in classifying three forest types. The following are the training field spectral statistics:

Forest Type	Mean Spectral Value	Standard Deviation
Black Spruce	20	10
Paper Birch	40	10
Quaking Aspen	50	10

Plot the likelihood curves for the tree cover classes on graph paper.

After determining the optimal cutoff values from your likelihood plot, classify the following image using the maximum likelihood rule. Color your classified image using the following color palette assignments:

Black Spruce → black

Paper Birch → green

Quaking Aspen → yellow

Original Image

21	23	24	31	37	39
22	24	27	31	35	41
23	25	26	32	34	42
29	33	33	34	35	36
32	33	37	44	48	49

Classified Image

6) A watershed hydrologist is interested in classifying snow cover within a drainage basin. From training fields the following likelihood contours are developed:

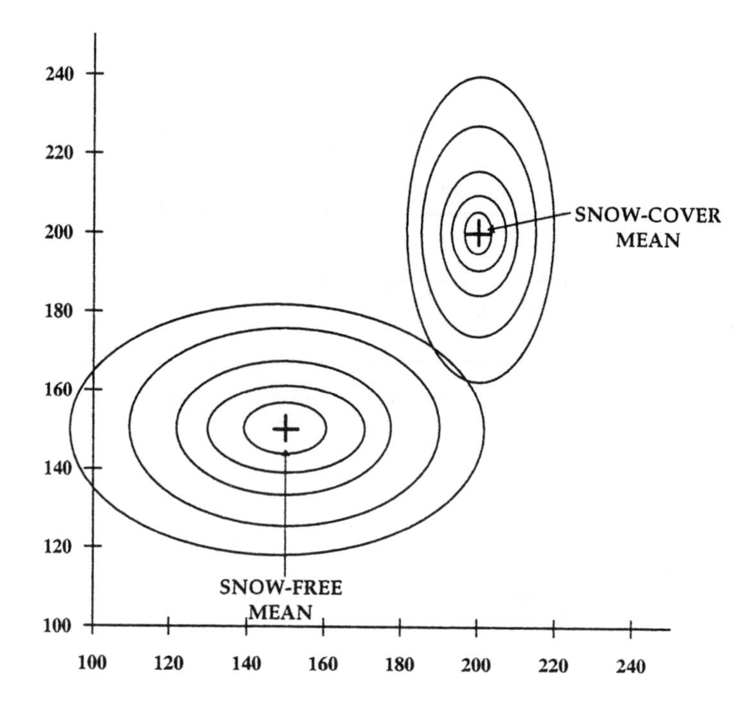

Using the maximum-likelihood rule, predict whether the following pixels are snow-covered or snow-free. Shade in the snow-free pixels.

180	180	190	190	190
155	160	160	180	185
180	180	192	190	190
156	157	162	165	190

ADDITIONAL READINGS

General

Bolstad, P. V. and T. M. Lillesand. 1992. Semi-automated training approaches for spectral class definition. *International Journal of Remote Sensing.* 13:3157–3166.

Campbell, J. B. 1981. Spatial correlation effects upon the accuracy of supervised classification of land cover. *Photogrammetric Engineering and Remote Sensing.* 47:355–363.

Foody, G. M., Campbell, N. A., Trodd, N. M. and T. F. Wood. 1992. Derivation and applications of class membership from the maximum-likelihood classification. *Photogrammetric Engineering and Remote Sensing.* 58:1335–1341.

Franklin, S. E. and B. A. Wilson. 1992. A three-stage classifier for remote sensing of mountain environments. *Photogrammetric Engineering and Remote Sensing.* 58:449–454.

Gong, P. and P. J. Howarth. 1990. An assessment of some factors influencing multispectral land-cover classification. *Photogrammetric Engineering and Remote Sensing.* 56:597–603.

Hixon, M. D., Scholz, D., Fuhs, N. and T. Akiyama. 1980. Evaluation of several schemes for classification of remotely sensed data. *Photogrammetric Engineering and Remote Sensing.* 46:1547–1553.

Kenk, E., Sondheim, M., and B. Yee. 1988. Methods for improving accuracy of Thematic Mapper ground cover classifications. *Canadian Journal of Remote Sensing.* 14:17–31.

Richards, J. A., Landgrebe, D. A. and P. H. Swain. 1982. A means of utilizing ancillary information in multispectral classification. *Remote Sensing of Environment.* 12:463–477.

Strahler, A. H. 1980. The use of prior probabilities in maximum likelihood classification. *Remote Sensing of Environment.* 10:135–163.

Applications

Chuvieco, E. and R. G. Congalton. 1988. Mapping and inventory of forest fires from digital processing of TM data. *Geocarto International.* 3:41–53.

Franklin, S. E. 1992. Satellite remote sensing of forest type and landcover in the subalpine forest region, Kananaskis Valley, Alberta. *Geocarto International.* 7:25–35.

Nel, E. M., Wessman, C. A., and T. T. Veblen. 1994. Digital and visual analysis of Thematic Mapper imagery for differentiating old growth from younger spruce-fir stands. *Remote Sensing of Environment.* 48:291–301.

Ringrose, S., Matheson, W., Tempest, F. and T. Boyle. 1990. The development and causes of range degradation features in southeast Botswana using multi-temporal Landsat MSS imagery. *Photogrammetric Engineering and Remote Sensing.* 56:1253–1262.

CHAPTER 8

Assessment of Classification Accuracy

INTRODUCTION

A common question asked by potential user's of a classified image is: "How accurate is the classification?" Visual inspection of a classified image can be misleading; an elk biologist may look at some favorite aspen stands and see that they were all correctly classified, while a grizzly bear biologist may see that many critical whitebark pine stands were misclassified. These two biologists would have very different conclusions about the accuracy of the classification. Fortunately, there has been significant research on classification accuracy assessment techniques over the last decade. This chapter covers the basics of classification accuracy assessment; for more details please read some of the articles in the *Additional Readings* section.

THE ERROR MATRIX

Let's look at a simple example. Imagine that you have classified an image from a small area in the Rocky Mountains. The major cover types within your study area are quaking aspen (QA), lodgepole pine (LP), Engelmann spruce/subalpine fir (SF), alpine meadow (AM), and water bodies (W). For each of the classified cover types you establish at least 30 random sample points. You then visit each random sample point in the field and verify it's actual cover type (recorded as reference data). You then produce a matrix, called an error matrix (sometimes termed confusion matrix or contingency table) with the reference data columns representing the actual "ground truth" from field verification of each random sample point. The rows of the error matrix represent the predicted classes for the random sample points (Table 8.1.). The overall classification accuracy can be computed as the total number of correct class predictions (the sum of the diagonal cells) divided by the total number of cells. In our example, the overall classification accuracy was (40 + 30 + 25 + 50 + 32) / 200, or 88%.

Table 8.1. Simple example of an error matrix

Predicted Cover Type	Reference Data ("Ground Truth")					Row Totals
	QA	LP	SF	AM	W	
QA	**40**	0	0	3	0	43
LP	0	**30**	12	0	1	43
SF	0	3	**25**	0	2	30
AM	2	0	0	**50**	0	52
W	0	0	0	0	**32**	32
Column totals	42	33	37	53	35	200

Table 8.2. Producer's and user's accuracy by cover type class

Cover Type	Producer's Accuracy	User's Accuracy
Quaking Aspen	40/42 * 100 = 95%	40/43 * 100 = 93%
Lodgepole Pine	30/33 * 100 = 91%	30/43 * 100 = 70%
Spruce/Fir	25/37 * 100 = 68%	25/30 * 100 = 83%
Alpine Meadow	50/53 * 100 = 94%	50/52 * 100 = 96%
Water	32/35 * 100 = 91%	32/32 * 100 = 100%

User's and Producer's Accuracy

We can evaluate the classification accuracy of each cover type in two different ways, termed *user's accuracy and producer's accuracy.* User's accuracy is the percentage of pixels that were predicted to be a cover type that actually were that cover type (as determined by "ground truth"). The user's accuracy of each cover type can be computed using the error matrix. For example, we predicted 43 sample points to be quaking aspen and 40 of those points were actually quaking aspen, while 3 points were actually alpine meadow. Therefore the user's accuracy for quaking aspen is 40/43 * 100 = 93%. This high accuracy might be due to the homogeneous nature of even-aged, single-story aspen stands in our area. Other more heterogeneous cover types might have poorer user's accuracy. For example, spruce/fir can be uneven-aged, and multi-storied and therefore might be more difficult to accurately classify. In our example, the user's accuracy for spruce/fir is 25/30 * 100 = 83%.

User's accuracy reflects commission errors; of the pixels committed (predicted) to a class, some predictions will be incorrect. Producer's accuracy reflects omission errors; of the points that actually belong to a class ("ground truth" determined in the field), some of the points will be incorrectly classified. For example, using our error matrix, there were 42 points that were determined in the field to be quaking aspen and 40 of these points were classified (predicted) to be quaking aspen. Therefore the producer's accuracy for quaking aspen is 40/42 * 100 = 95%. Producer's accuracy and user's accuracy can be quite different for a class. For example, the producer's accuracy for spruce/fir was 68%, while the user's accuracy was 83% (Table 8.2). Therefore both the producer's and user's accuracy for each class should be reported with the error matrix.

Kappa Statistic

How much better is our classification compared to one where we randomly assigned class values to each pixel? The kappa statistic (KHAT) can be used to answer this question. KHAT can be computed as:

$$\text{KHAT} = \frac{\text{Overall Classification Accuracy} - \text{Expected Classification Accuracy}}{1 - \text{Expected Classification Accuracy}}$$

Table 8.3. Matrix of products for calculating the kappa statistic value (KHAT)

	QA	LP	SF	AM	W	Table 8.1 Row Totals
QA	**1806**	1419	1591	2279	1505	43
LP	1806	**1419**	1591	2279	1505	43
SF	1260	990	**1110**	1590	1050	30
AM	2184	1716	1924	**2756**	1820	52
W	1344	1056	1184	1696	**1120**	32
Table 8.1 column totals	42	33	37	53	35	200

The expected classification accuracy is the accuracy expected based on chance, or the expected accuracy if we randomly assigned class values to each pixel. It can be calculated by first using the error matrix to produce a matrix of products of row and column totals (Table 8.3). For example, the first cell of the matrix is the product of row #1 and column #1 totals from the error matrix (43 * 42 = 1806).

The expected classification accuracy is then computed as the sum of the diagonal cell values divided by the sum of all cell values. In our example, the expected classification accuracy is:

$$(1806 + 1419 + 1110 + 2756 + 1120) / 40{,}000 * 100 = 21\%$$

Therefore the KHAT value for our classification is:

$$(0.88 - 0.21) / (1 - 0.21) = 0.85$$

Based on the KHAT value, we can state that our classification is 85% better than that expected if we randomly assigned a cover type class to each image pixel.

COLLECTION OF REFERENCE DATA

A crucial assumption in producing the error matrix, user's and producer's accuracy values, and the KHAT value, is that the reference data are truly representative of the entire classification. It is possible to produce a misleading assessment of classification accuracy depending on how the reference data are collected.

Sources of Conservative Estimates of Classification Accuracy

The actual accuracy of our classification is unknown because it is impossible to perfectly assess the true class of every pixel. However, we can use a sample of random points to build an error matrix for estimating the accuracy of our classification. If our estimate is less than the actual classification accuracy, we have made a conservative estimate. Let's take an extreme example. Imagine that we have a perfect classification — every pixel is correctly classified, and therefore, our actual overall classification accuracy is 100%.

Errors in Reference Data

Our estimate of our overall classification accuracy will probably be less than 100% for several reasons. First, we assume that our "ground truth" data are perfect. However, if there are any errors in our reference data such as incorrect class assignment, change in covertype between the time of imaging and the time of field verification, mistakes in recording or processing the reference data, some of our correctly classified pixels may be incorrectly assessed as being misclassified.

Positional Errors

Because we rectified the image with a certain tolerance of positional error, we can not be absolutely sure of the location of any given pixel. For example, if we rectified an image to 30-meter output pixels with an RMS error of 1.0, the average positional error of the model is ±30 meters. Some pixels will have positional accuracy better than this and some pixels will have positional accuracy that is worse than this. Therefore, even if we use submeter GPS technology to navigate us to the center coordinates of a pixel to be "ground truthed," we cannot be certain that we are in that pixel. Because of the positional error inherent in rectified images, some correctly classified pixels may not be correctly located during field sampling. This problem of positional error will also lead to a conservative estimate of classification accuracy (Figure 8.1).

Minimum Mapping Unit Area

Sometimes aerial photography is interpreted as a substitute for field collection of reference data. Because aerial photography is visually interpreted as polygons, the minimum mapping unit area used in the interpretation can also lead to conservative estimates of classification accuracy. For example, if one interprets cover types to a minimum mapping unit area of 1 hectare, a classified image of 30-meter pixels (0.09 ha) may contain many small clusters of single-class pixels that were too small to be included in the interpretation of the aerial photography. This will lead to a conservative estimate of classification accuracy (Figure 8.2).

Sources of Optimistic Estimates of Classification Accuracy

It is possible to have a poor classification and yet report a high classification accuracy estimate. There are several potential sources of this optimistic bias.

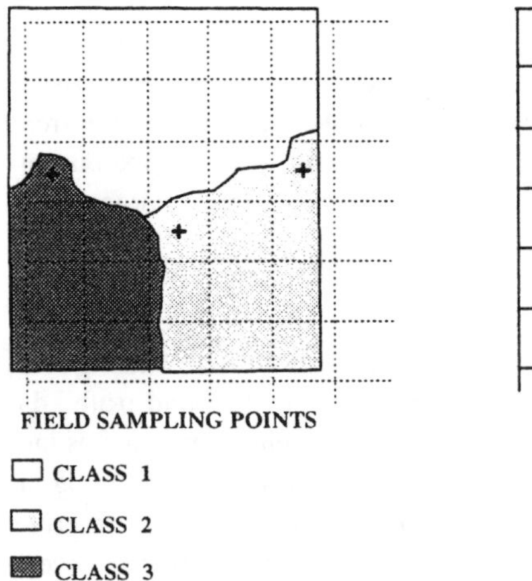

1	1	1	1	1
1	1	1	1	1
1*	1	1	1	2*
3	3	2*	2	2
3	3	2	2	2
3	3	2	2	2

CLASSIFIED IMAGE

Figure 8.1. Conservative estimate of classification accuracy due to positional error. Pixels with * have been randomly selected for field validation. Row 3, Column 1 is correctly classified but assessed as being incorrect due to slight positional error.

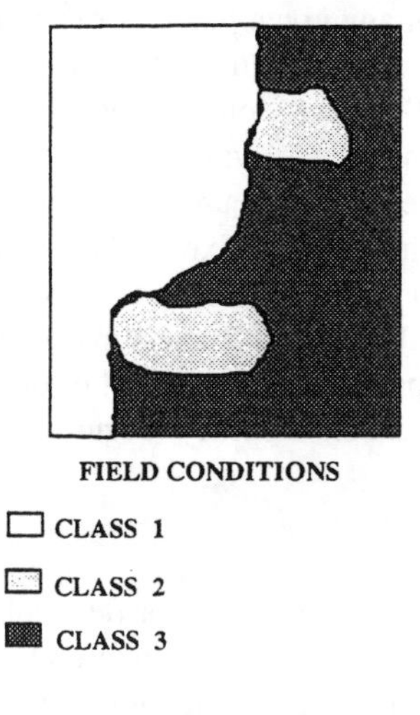

1	1	1	3	3
1	1	1	2*	3
1	1	1	3	3
1	1	3	3	3
1	2*	2*	3	3
1	3	3	3	3

CLASSIFIED IMAGE

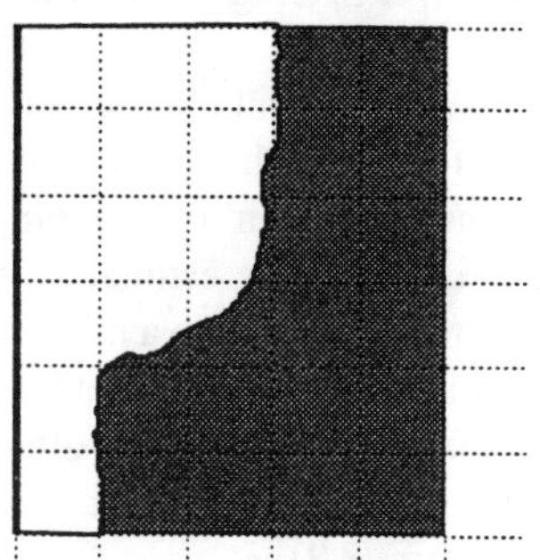

Figure 8.2. Conservative estimate of classification accuracy due to large minimum mapping unit of reference data.

Training Fields as Reference Data

It is tempting to minimize cost and time by using the data collected from the training field areas as reference data. This can lead to optimistic estimates of classification accuracy for at least three reasons. First, the pixels selected for training field data are usually selected because they are from relatively homogeneous areas — areas that are typically easier to correctly classify compared to other areas in the image. For example, we might estimate our overall classification accuracy to be 100% if we selected our reference data from training fields that encompassed large, pure, level stands. However, if much of our image is a heterogeneous mixture of vegetation types of variable sizes, density and topographic conditions, our classified image is probably not really 100% accurate. Second, if we collect reference data near our training field areas, the cover types encountered are likely to belong to one of our vegetation classes. However, if we randomly located plots for "ground truthing," we might find that some points belong to a vegetation class that was not originally included in the classification. Third, because we are developing a statistical model with the training field data, estimates of classification accuracy are likely to be optimistically biased because we use the same data for model development and model validation.

Sampling from Blocks of Classified Pixels

Because of the positional error inherent in rectified images, a common strategy is go to the coordinates of the center pixel of a 3 by 3 group of pixels belonging to the same class (Figure 8.3). However, this can lead to an optimistic estimate of classification accuracy since homogeneous areas tend to be selected while heterogeneous areas that are more difficult to correctly classify are excluded from selection as reference data.

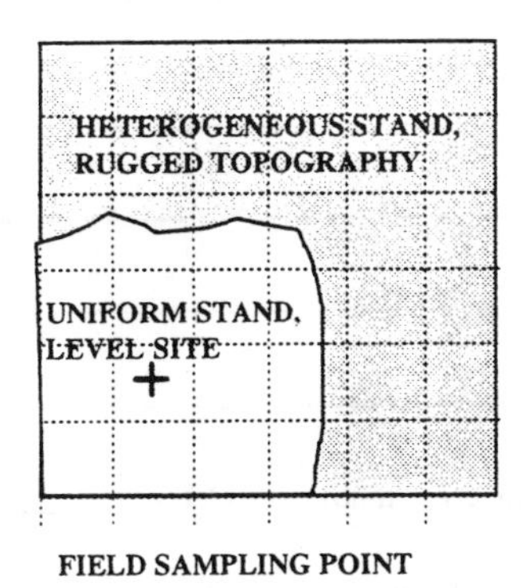

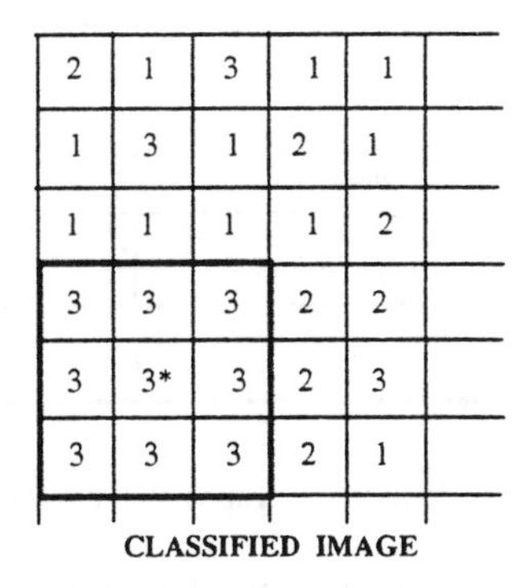

Figure 8.3. Example of optimistic bias in assessment of classification accuracy due to restricting sampling to blocks of single-class pixels.

The error matrix, producer's and user's accuracy and KHAT value have become standard in assessment of classification accuracy. However, if the error matrix is improperly generated by poor reference data collection methods, then the assessment can be misleading. Therefore sampling methods used for reference data should be reported in detail so that potential users can judge whether there may be significant biases in the classification accuracy assessment.

CHAPTER 8 PROBLEMS

1) A forester classifies an image into the following classes: bottomland hardwoods (BH), slash pine (SP), loblolly pine (LP), and water (W). One thousand random points are then established and visited for classification accuracy assessment as follows:

"GROUND TRUTH"

CLASSIFIED	BH	SP	LP	W
BH	69	11	56	15
SP	4	121	14	5
LP	6	71	468	6
W	9	0	0	145

Determine the user's and producer's accuracy for each of the classes:

Cover Type	Producer's Accuracy	User's Accuracy
Bottomland Hardwoods		
Slash Pine		
Loblolly Pine		
Water		

Calculate the Kappa statistic value for this error matrix: KHAT = _______.

2) Give an example of an error matrix that would have a KHAT value of 0.0.

"GROUND TRUTH"

CLASSIFIED	BH	SP	LP	W
BH				
SP				
LP				
W				

3) A range manager wants to evaluate a digital classification of three dominant range types: sagebrush, cheatgrass, and shadscale. One thousand points are randomly located throughout the area for accuracy assessment. However, because of budget and time constraints, only those random sample points to within 1/2 mile of existing roads are visited. At least 100 random points from each cover class are sampled for classification accuracy assessment.

   The overall classification accuracy was 85%. Is this estimate likely to be conservative or optimistic?

   Explain your reasoning.

4) A wetlands scientist obtains a Landsat Thematic Mapper classification of a 10,000 hectare freshwater marsh. To assess the accuracy of the digital classification, an accurate vector GIS of ten randomly located 100-ha blocks is developed by interpreting color infrared aerial photographs taken the same day the Landsat TM image was acquired. The minimum mapping unit for photo interpretation was 2.5 acres (smallest polygon is 2.5 acres).

   The overall classification accuracy of the digital classification was assessed as 85% using the GIS as reference data. Is this estimate likely to be conservative or optimistic?

   Explain your reasoning.

5) A forest products company obtains digital classification of forest covertypes. To assess the accuracy of their digital classification, they randomly select 3-column by 3-row groups of pixels that have the same class. These are selected as reference sample units. They then use submeter GPS technology to navigate to the center of each reference sample unit. At least 100 sample units from each cover class are ground checked.

   The overall classification accuracy of the digital classification was assessed as 85% using the GIS as reference data. Is this estimate likely to be conservative or optimistic?

   Explain your reasoning.

6) A forester produces a digital classification of old growth forest. The accuracy of the classification is assessed independently by several groups. Match the following accuracy assessment scenarios with the most likely overall classification accuracy.

____ 50% overall classification accuracy
____ 75% overall classification accuracy
____ 90% overall classification accuracy

Group A: A guided approach was conducted in conjunction with the field work for training data collection, thereby significantly reducing the costs of data collection. A systematic sample in which the center of a polygon from the center of every 20th air photo plot was used as reference samples. To minimize nonclassification errors (such as slight positional shifts) only homogeneous polygons of at least 5 hectares were used as reference data. The center pixel of these polygons was selected for each reference data sample location. The photos were taken the same month and year as the satellite image was acquired.

Group B: Using high altitude color infrared photographs (taken by NASA in 1979), we stereoscopically interpreted cover types to polygons with a minimum mapping unit of 2 hectares. The covertypes were transferred to the USGS 1:63,360 quadrangle series using a digitizing tablet and stored in a vector GIS. For accuracy assessment these GIS polygons were rasterized to correspond to the classified image. Classification accuracy assessment was conducted on a pixel by pixel basis between the reference data and the classified image for a test area of 1000 ha.

Group C: All pixels that were classified as old growth were selected. A random sample of 100 classified old growth pixels then selected. For each selected pixel, we recorded the pixel UTM coordinates using a cell coordinate inquiry. We then programmed a 6-channel GPS to guide us to these UTM coordinates. The actual covertype was then recorded at these GPS-guided locations.

ADDITIONAL READINGS

Arnoff, S. 1982. Classification accuracy: a user approach. *Photogrammetric Engineering and Remote Sensing.* 48:1299–1307.

Arnoff, S. 1982. The map accuracy report: a user's view. *Photogrammetric Engineering and Remote Sensing.* 48:1309–1312.

Arnoff, S. 1983. Evaluating the effectiveness of remote sensing derived data for environmental planning. *Journal of Environmental Management. 17*:277–290.

Arnoff, S. 1985. The minimum accuracy value as an index of classification accuracy. *Photogrammetric Engineering and Remote Sensing.* 51:99–111.

Card, D. H. 1982. Using known map category marginal frequencies to improve estimates of thematic map accuracy. *Photogrammetric Engineering and Remote Sensing.* 48:431–439.

Card, D. H. 1989. Accuracy assessment, using stratified plurality sampling of portions of a landsat classification of the Arctic National Wildlife Refuge coastal plain. *NASA Technical Memorandum 101042*. Ames Research Center, Moffett Field, CA. 51 pp.

Congalton, R. G. and R. A. Mead. 1983. A quantitative method to test for consistency and correctness in photointerpretation. *Photogrammetric Engineering and Remote Sensing*. 49:69–74.

Congalton, R. G. 1988. A comparison of sampling schemes used in generating error matrices for assessing the accuracy of maps generated from remotely sensed data. *Photogrammetric Engineering and Remote Sensing*. 54:593–600.

Congalton, R. G. 1988. Using spatial autocorrelation analysis to explore errors in maps generated from remotely sensed data. *Photogrammetric Engineering and Remote Sensing*. 54:587–592.

Congalton, R. G. 1991. A review of assessing the accuracy of classifications of remotely sensed data. *Remote Sensing of Environment*. 37:35–46.

Congalton, R. G. and G. S. Biging. 1992. A pilot study evaluating ground reference data collection efforts for use in forest inventory. *Photogrammetric Engineering and Remote Sensing*. 58:1669–1671.

Dicks, S. E., and T. H. C. Lo. 1990. Evaluation of thematic map accuracy in a land-use and land-cover mapping program. *Photogrammetric Engineering and Remote Sensing*. 56:1247–1252.

Fitzgerald, R. W. and B. G. Lees. 1994. Assessing the classification accuracy of multisource remote sensing data. *Remote Sensing of Environment*. 47:362–368.

Fitzpatrick-Lins, K. 1981. Comparison of sampling procedures and data analysis for a land-use and land-cover map. *Photogrammetric Engineering and Remote Sensing*. 47:343–351.

Foody, G. M. 1988. Incorporating remotely sensed data into a GIS: the problem of classification evaluation. *Geocarto International*. 3:13–16.

Foody, G. M. 1992. On the compensation for chance agreement in image classification accuracy assessment. *Photogrammetric Engineering and Remote Sensing*. 58:1459–1460.

Guinevan, M. E. 1979. Testing land-use map accuracy: another look. *Photogrammetric Engineering and Remote Sensing*. 45:1371–1377.

Hay, A. M. 1979. Sampling designs to test land-use map accuracy. *Photogrammetric Engineering and Remote Sensing*. 45:529–533.

Hay, A. M. 1988. The derivation of global estimates from a confusion matrix. *International Journal of Remote Sensing*. 9:1395–1398.

Hord, R. M. and W. Brooner. 1976. Land use map accuracy criteria. *Photogrammetric Engineering and Remote Sensing*. 42:671–677.

Hudson, W. D. and C. W. Ramm. 1987. Correct formulation of the Kappa coefficient of agreement. *Photogrammetric Engineering and Remote Sensing*. 53:421–422.

Lunetta, R. S., Congalton, R. G., Fenstermaker, L. K., Jensen, J. R., McGwire, K. C. and L. R. Tinney. 1991. Remote sensing and geographic information system data integration: error sources and research issues. *Photogrammetric Engineering and Remote Sensing*. 57:677–687.

Martin, L. R. G. 1989. Accuracy assessment of landsat-based change detection methods applied to the rural-urban fringe. *Photogrammetric Engineering and Remote Sensing*. 55:209–215.

Maxim, L. D., Harrington, L. and M. Kennedy. 1981. Alternative scale-up estimates for aerial surveys where both detection and classification error exist. *Photogrammetric Engineering and Remote Sensing.* 47:1227–1239.

Mead, R. A. and J. Szajgin. 1982. Landsat classification accuracy assessment procedures. *Photogrammetric Engineering and Remote Sensing.* 139–141.

Prisley, S. P. and J. Smith. 1987. Using classification error matrices to improve the accuracy of weighted land-cover models. *Photogrammetric Engineering and Remote Sensing.* 53:1259–1263.

Rosenfield, G. H., Fitzpatrick-Lins, K. and H. S. Ling. 1982. Sampling for thematic map accuracy testing. *Photogrammetric Engineering and Remote Sensing.* 48:131–137.

Rosenfield, G. H. and K. Fitzpatrick-Lins. 1986. A coefficient of agreement as a measure of thematic classification accuracy. *Photogrammetric Engineering and Remote Sensing.* 52: 223–227.

Skidmore, A. K. and B. J. Turner. 1992. Map accuracy assessment using line intersect sampling. *Photogrammetric Engineering and Remote Sensing.* 58:1453–1457.

Star, J. L. 1989. Sources of errors in thematic classification of remotely sensed imagery. *Proceedings of IGARSS 89/12th Canadian Symposium on Remote Sensing.* Vancouver, BC. pp. 1851–1853.

Stehman, S. V. 1992. Comparison of systematic and random sampling for estimating the accuracy of maps generated from remotely sensed data. *Photogrammetric Engineering and Remote Sensing.* 58:1343–1350.

Thomas, I. L. and G. M. Allcock. 1984. Determining the confidence level for a classification. *Photogrammetric Engineering and Remote Sensing.* 50:1491–1496.

Todd, W. J., Gehring, D. G., and J. F. Haman. Landsat wildland mapping accuracy. *Photogrammetric Engineering and Remote Sensing.* 46(4):590–620.

van Genderen, J. L., Lock, B. F., and P. A. Vass. 1978. Remote sensing: statistical testing of thematic map accuracy. *Remote Sensing of Environment.* 7:3–14.

Wang, M. and P. J. Howarth. 1993. Modeling errors in remote sensing image classification. *Remote Sensing of Environment.* 45:261–271.

Appendix: Solutions to Even-Numbered Problems

CHAPTER 1

2) Suppose you have a Landsat Thematic Mapper image consisting of 1000 rows by 1000 columns. If you printed this image on a piece of paper that was 100 cm by 100 cm in size, what would the scale of the printed image be? **Assuming the image has been resampled to 30-meter pixels:**

   on the map 1 pixel = 100 cm/1000 columns = 0.1 cm wide
   0.1 cm on map / 30 m on ground (100 cm/m) = 0.0000333 or 1:30,000

4) Match the following resolution terms with the best analogy (list one letter for each resolution term):

   ____ Higher radiometric resolution **F) Using a meat thermometer**
   ____ Lower radiometric resolution **E) Using a pop-up indicator**
   ____ Higher spatial resolution **H) Taking a photograph with a 400 mm zoom lens**
   ____ Lower spatial resolution **G) Taking a photograph with a 25 mm wide angle lens**
   ____ Higher spectral resolution **B) Taking a color photograph with a yellow (minus blue) filter**
   ____ Lower spectral resolution **A) Taking a color photograph without using a filter**
   ____ Higher temporal resolution **C) Subscribing to a daily newspaper**
   ____ Lower temporal resolution **D) Subscribing to a monthly magazine**

CHAPTER 2

2) Given the following data on 9-track tape, fill in the proper values for each pixel in the image:

9-track tape information:
Data packaged as BIL (band interleaved by line)
Number of rows: 2 Number of columns: 2 Number of bands: 3

Tape File:

Integer Value:	Bit#8 (2^7) **=128**	Bit#7 (2^6) **=64**	Bit#6 (2^5) **=32**	Bit#5 (2^4) **=16**	Bit#4 (2^3) **=8**	Bit#3 (2^2) **=4**	Bit#2 (2^1) **=2**	Bit#1 (2^0) **=1**
21	0	0	0	1	0	1	0	1
26	0	0	0	1	1	0	1	0
32	0	0	1	0	0	0	0	0
28	0	0	0	1	1	1	0	0
47	0	0	1	0	1	1	1	1
57	0	0	1	1	1	0	0	1
59	0	0	1	1	1	0	1	1
54	0	0	1	1	0	1	1	0
67	0	1	0	0	0	0	1	1
80	0	1	0	1	0	0	0	0
47	0	0	1	0	1	1	1	1
57	0	0	1	1	1	0	0	1

IMAGE:

BAND 1:	21	26
BAND 2:	32	28
BAND 3:	47	57
BAND 1:	59	54
BAND 2:	67	80
BAND 3:	47	57

4) Match the following colors with the appropriate RGB video intensities:

COLOR CHOICES:	DISPLAYED COLOR:	RED VIDEO INTENSITY	GREEN VIDEO INTENSITY	BLUE VIDEO INTENSITY
	4) White	255	255	255
	6) Gray	125	125	125
	5) Black	0	0	0
	1) Red	255	0	0
		255	150	0
	2) Green	0	255	0
	3) Blue	0	0	255
	9) Yellow	255	255	0
	13) Navy Blue	0	0	70
10) Brown		65	255	130
11) Sand		150	150	0
12) Orange	**7) Cyan**	0	255	255
	8) Magenta	255	0	255
14) Mint Green		215	167	116
15) Avocado		95	60	0

The easiest choices have been made. The remaining choices are brown, sand, orange, and mint green, avocado. The remaining colors in red, green, blue are:

255,150,0 — red is 255,0,0 and yellow is 255,255,0.
orange is halfway between these colors — orange
95,60,0 — a very dark orange — brown
65,255,130 — a bright green, with moderate blue — mint green
215,167,116 — bright red, moderate green, less blue — sand
150,150,0 — yellow is 255,255,0 so this would be a dark yellow — avocado

6) You have a computer monitor that displays red, green, and blue in 0–255 video intensities. You also have a color ink-jet printer that prints with cyan, magenta, and yellow inks in 0–15 shades. Write three functions that will convert any red, green, and blue video intensities to the proper cyan, magenta, and yellow print shades such that a color image printed out will appear similar to the color image displayed on the computer monitor screen.

   Cyan = **(255 – Red Value) / 255 * 15**
   Magenta = **(255 – Green Value) / 255 * 15**
   Yellow = **(255 – Blue Value) / 255 * 15**

   Examples:

   on screen: 255 red, 255 green, 0 blue — bright yellow video display = 0 cyan, 0 magenta 15 yellow — bright yellow print shade

   on screen: 0 red, 255 green, 255 blue — bright cyan video display = 15 cyan, 0 magenta, 0 yellow — bright cyan print shade

   on screen: 100 red, 100 green 100 blue — gray video display = 9 cyan, 9 magenta, 9 yellow — gray print shade

   on screen: 255 red, 255 green, 255 blue — white video display = 0 cyan, 0 magenta, 0 yellow — assuming paper is white — white print shade

8) What term does the following 3 bytes represent using ASCII coding?

0	1	0	0	0	1	1	1

0	1	0	0	1	0	0	1

0	1	0	1	0	0	1	1

$2^6 + 2^2 + 2^1 + 2^0$
$= 64 + 4 + 2 + 1$
$= 71$
ASCII code 71
= G

$2^6 + 2^3 + 2^0$
$= 64 + 8 + 1$
$= 73$
ASCII code 73
= I

$2^6 + 2^4 + 2^1 + 2^0$
$= 64 + 16 + 2 + 1$
$= 83$
ASCII code 83
= S

10) Using the 10-row by 10-column image from problem 9, develop a min/max linear stretch. What are the pixel video intensity values after the linear stretch enhancement?

   From the image: minimum = 5, maximum = 32

   Linear Stretch = (Digital Value – Min) / (Max – Min) * 255

   = (Digital Value – 5) / 27 * 255

Lookup Table:

Pixel Value	Video Intensity	Pixel Value	Video Intensity
0	**0**	17	**113**
1	**0**	18	**123**
2	**0**	19	**132**
3	**0**	20	**142**
4	**0**	21	**151**
5	**0**	22	**161**
6	**9**	23	**170**
7	**19**	24	**179**
8	**28**	25	**189**
9	**38**	26	**198**
10	**47**	27	**208**
11	**57**	28	**217**
12	**66**	29	**227**
13	**76**	30	**236**
14	**85**	31	**246**
15	**94**	32	**255**
16	**104**	33	**255**

Displayed Video Intensities:

0	28	85	85	85	47	38	19	19	28
0	19	85	76	76	47	38	9	19	19
9	19	57	66	76	57	47	9	9	9
9	28	38	57	66	47	47	9	0	9
114	123	142	95	85	76	47	9	0	9
133	142	161	142	133	133	66	9	9	9
198	208	114	133	133	133	85	28	38	38
227	217	104	142	133	133	114	47	47	38
255	236	152	152	142	152	114	57	47	47
246	236	198	189	161	161	95	66	47	47

CHAPTER 3

2) Color infrared aerial photography is often used in natural resource management for vegetation mapping. The color infrared film is exposed to wavelengths from 0.5 to 0.9 mm. List any satellite systems that have spectral bands that entirely cover this spectral region.

Landsat Thematic Mapper (TM band 2: 0.52 mm – 0.60 μm TM band 3: 0.63 – 0.69 μm, TM band 4: 0.76 – 0.90 μm)

Landsat Multispectral Scanner (MSS band 1: 0.5 – 0.6 μm, MSS band 2: 0.6 – 0.7 μm, MSS band 3: 0.7 – 0.8 μm, MSS band 4: 0.8 – 1.1 μm)

SPOT HRV XS (SPOT HRV XS band 1: 0.50 – 0.59 μm, SPOT HRV XS band 2: 0.61 – 0.68 μm, SPOT HRV XS band 3: 0.79 – 0.89 μm)

AVHRR (AVHRR band 1: 0.58 – 0.68 μm, AVHRR band 2: 0.73 – 1.10 μm)

4) Suppose the following linear regression model was developed to predict surface water temperature in a large lake.

$$\text{TEMPERATURE} = -42.5 + 0.5 * (\text{DIGITAL VALUE})$$

Using the following thermal digital image, predict the surface water temperature for each grid cell.

133	130	129	129	128	127	125	123	120	118
131	129	128	128	127	128	126	122	119	116
129	129	128	128	127	125	122	119	118	115
127	126	125	125	125	125	123	119	117	113

24	22.5	22	22	21.5	21	20	19	17.5	16.5
23	22	21.5	21.5	21	21.5	20.5	18.5	17	15.5
22	22	21.5	21.5	21	20	18.5	17	16.5	15
21	20.5	20	20	20	20	19	17	16	14

6) **There is a strong linear relationship between leaf area and the ratio of near-infrared/red reflectance. For example read:**

Running, S. W., Peterson, D. L., Spanner, M. A. and K. B. Teuber. 1986. Remote sensing of coniferous forest leaf area. *Ecology.* 67:273–276.

Then read:

Spanner, M. A., Pierce, L. L., Peterson, D. L. and S. W. Running. 1990. Remote sensing of temperate coniferous forest leaf area index: the influence of canopy closure, understory vegetation, and background reflectance. *International Journal of Remote Sensing.* 11:95–111.

There was a strong relationship between near infrared reflectance and leaf area index of closed stands. Assume we developed a linear equation using this relationship. If we applied the model to the following stands, would we underestimate or overestimate leaf area index? Why?

a) An old growth conifer stand with large trees and an understory of forest litter. **Underestimate because with high leaf area index the reflectance would be reduced due to intercrown shadowing.**

b) An open canopy conifer stand with an understory of broadleaf shrubs or grass. **Overestimate. The leaf area index would be relatively low, yet reflectance would be relatively high due to the broadleafs or grass understory.**

c) An open canopy conifer stand with an understory of granite bedrock. **Overestimate. The leaf area index would be relatively low, yet reflectance would be relatively high due to the dry bedrock and dry granitic soil.**

d) A mixed conifer/broadleaf stand (75% conifer, 25% broadleaf). **Overestimate. The leaf area index would be relatively low, yet reflectance would be relatively high due to the broadleafs.**

8) Read the following:
Lathrop, R. G. Jr. 1992. Landsat Thematic Mapper monitoring of turbid inland water quality. *Photogrammetric Engineering and Remote Sensing.* 58: 465–470.

Why is it difficult to estimate phytoplankton chlorophyll concentration from satellite data of turbid inland lakes? **Because the suspended solids are relatively highly reflective and therefore spectral reflectance due to chlorophyll concentration is masked by reflectance due to suspended solids.**

Why is it risky to extrapolate an existing remote sensing water quality model to regions that are different than the region where the model was developed? **Because the empirical model may be site-specific and only applicable to the lakes were the data were collected from. Application of such models to regions with different geology, vegetation, or land use may produce misleading results.**

CHAPTER 4

2) Read the paper by Robinove (1982) on computing physical values from digital numbers. Then list one periodical article that actually used physical values rather than digital numbers.
List the complete citation from each paper including all authors, date, periodical, volume and pages. Discuss why your selected study used physical values instead of digital numbers.

Badhwar, G. D., MacDonald, R. B. and N. C. Mehta. 1986. Satellite-derived leaf-area-index and vegetation maps as input to global carbon cycle models — a hierarchical approach. *International Journal of Remote Sensing*. 7:265–281. The authors converted Landsat digital values to absolute reflectance so that they could compare these data with aircraft and helicopter data.

Spanner, M. A., Pierce, L. L., Peterson, D. L., and S. W. Running. 1990. Remote sensing of temperate coniferous forest leaf area index: The influence of canopy closure, understory vegetation and background reflectance. *International Journal of Remote Sensing*. 11:95–111. The authors converted Landsat Thematic Mapper digital values to radiance so that the results could be compared with ground-based radiometer measurements.

Pitblado, J. R. 1992. Landsat views of Sudbury (Canada) area acidic and nonacidic lakes. *Canadian Journal of Fisheries and Aquatic Sciences*. 49:33–39. Landsat Thematic Mapper and MSS digital values were converted to radiances to enable comparisons between these two sensors.

CHAPTER 5

2) If state plane coordinate tics are printed every 10,000 ground-feet along the borders of your 1:24,000 map, what is the spacing interval in map inches for these tic marks? Show your calculations.

(10,000 ft on ground) (1 map unit / 24,000 ground units) (12 inches/ 1 foot) = 5 inches on map

4) Why are most USGS 7.5 minute quadrangles not square?

   Because the quadrangles are 7.5 minutes in longitude and latitude — a spherical or nonplanar coordinates system.

   Can you think of any examples where you would expect a quadrangle to be square?

   Probably not because a unit of longitude is not the same linear distance as a unit of latitude, even at the equator.

6) Some people recommend rectifying an image after it has been classified. Their rational usually is 1) Rectifying is quicker since each pixel contains only one class value instead of many spectral values, 2) Some spectral integrity is lost during the pixel resampling process — an unrectified image is spectrally more correct than a rectified image and is therefore preferred for image classification. List at least two reasons why you might want to rectify an image before classifying it for vegetation mapping.

   Reason #1: **If you rectify before classification, then you could incorporate planimetric ancillary data such as elevation, slope, aspect, and soils in the classification.**

   Reason #2: **With a rectified image, you could incorporate GIS or GPS-based information for guidance in delineating training fields.**

   Reason #3: **Rectification is done only once. It is probably more efficient to rectify the image once as a preprocessing step rather than rectifying after each of many different classifications.**

   Reason #4: **The analyst becomes more familiar with the image prior to classification by selecting ground control points across the image. This can only help the analyst in making subsequent training field and classification decisions.**

   If you rectified *after* you classify, you would be able to use ONLY one resampling method — nearest neighbor resampling. For example, if 1 = aspen/birch, 2 = spruce, 3 = willow, etc., rectification after classification using bilinear interpolation or cubic convolution resampling would result in non-integer classes. For example, what would a class of 1.489 represent? Since you have no choice — you must use nearest neighbor resampling to rectify a classified image, rectify the image before classification and use nearest neighbor resampling if you are concerned about a "spectrally correct" rectified image.

8) The following transformation models were developed using a Landsat Thematic Mapper image of coastal area. After throwing out the poorest ground control pixels, seven ground control pixels remained.

GCP#	X pixel	X UTM	Y pixel	Y UTM
1	191	598,285	180	3,627,280
2	98	595,650	179	3,627,730
3	137	596,250	293	3,624,380
4	318	602,200	115	3,628,530
5	248	600,350	83	3,629,730
6	255	600,440	113	3,628,860
7	272	600,540	196	3,626,450

The transformation models are:

X pixel = -382.1 + 0.0341877 (Xmap) – 0.0054810(Ymap)
Y pixel = 130,163 – 0.00557618(Xmap) – 0.0349150(Ymap)

Suppose you use these models to rectify the Landsat image to 30-meter grid cells for on-screen digitizing in a Geographic Information System. What is the expected positional accuracy (in meters) of the rectified image. Show your work!

From the seven GCPs that were used to build the transformation models:

Point #	**Actual X**	**Predicted X**	**Residual X Error**	**Actual Y**	**Predicted Y**	**Residual Y Error**	**Root Squared Error**
1	191	**190.77**	**0.23**	180	**180.42**	**-0.24**	**0.33**
2	98	**98.22**	**-0.22**	179	**179.40**	**-0.40**	**0.46**
3	137	**137.09**	**-0.09**	293	**293.02**	**-0.02**	**0.09**
4	318	**317.76**	**0.24**	115	**114.95**	**0.05**	**0.25**
5	248	**247.94**	**0.06**	83	**83.37**	**-0.37**	**0.38**
6	255	**255.78**	**-0.78**	113	**113.24**	**-0.24**	**0.82**
7	272	**272.41**	**-0.41**	196	**196.83**	**-0.83**	**0.93**

Root Mean Squared Error = (0.33 + 0.46 + 0.09 + 0.25 + 0.38 + 0.82 + 0.93) / 7 = 0.47

10) Find one study that used each of the three most popular resampling methods: nearest neighbor, bilinear interpolation, and cubic convolution. For each

study, discuss whether you feel the most appropriate resampling method was used and why?

Nearest neighbor resampling: Rutchey, K. and L. Vilchek. 1994. Development of an Everglades vegetation map using a SPOT image and the global positioning system. *Photogrammetric Engineering and Remote Sensing.* 60:767–775. The authors rectified after classification, therefore they had no choice — they had to use nearest neighbor resampling.

Bilinear interpolation resampling: Novak, K. 1992. Rectification of digital imagery. *Photogrammetric Engineering and Remote Sensing.* 58:339–344. The author states that for most digital orthophoto applications, bilinear interpolation yields sufficiently good results. Cubic convolution would give better contrast but would take lots of computer time with large image files such as typical digital orthophotos. Nearest neighbor resampling would cause staircase patterns in linear features.

Cubic convolution resampling: Welch, R. and M. Ehlers. 1987. Merging multiresolution SPOT HRV and Landsat TM data. *Photogrammetric Engineering and Remote Sensing.* 53:301–303. The authors used cubic convolution resampling to avoid blocky image structures when merging the SPOT and Landsat images.

12) Suppose you were attempting to use the junction of streams within the Salmon River canyons as ground control locations. Which one of the following methods would you use and why?

____ geocoded image data
____ global positioning systems (GPS)
____ map interpolation using an engineers scale
____ map interpolation using a digitizing tablet

It would be difficult to receive at least 3 GPS satellite signals because of the steep canyon walls. Map interpolation with a digitizing tablet would be relatively inexpensive and quick — probably the best choice. However, the junction of streams may change location, so be careful if the map and image dates differ by 10 or 20 years.

CHAPTER 6

2) Using the image from question 1, perform the same sequential clustering, this time use only bands 1 and 2.

ORIGINAL IMAGE (BANDS 1 AND 2)

5,3	7,4	210,150	198,173
6,4	8,4	55,123	59,129
55,34	92,19	39,29	253,198

Pass #1: Pixel 1,1 starts as spectral class #1 with means of 5,3. Pixel 2,1 is close enough (<30) to belong to spectral class #1. Spectral class #1 has new means of 6,3.5. Pixel 3,1 is too far away, so it is assigned to a new spectral class — spectral class #2 with means of 210, 150. Pixel 4,1 is close enough (25.9) to belong to spectral class #2. Spectral class #2 therefore has two members with means of 204,161.5. Pixel 1,2 is close enough to belong to spectral class #1. Spectral class #1 has three members with means of 6,3.67. Pixel 2,2 is close enough to belong to spectral class #1. Spectral class #1 has four members with means of 6.5,3.75. Pixel 3,2 is too far from spectral class #1 or spectral class #2, and therefore we create a new spectral class #3 with means 55,123. Pixel 4,2 is close enough to belong to spectral class #3. Spectral class #3 has two members with means of 57,126. Pixel 1,3 is too far from any existing spectral class means and therefore becomes the first member of spectral class #4 with means of 55,34. Pixel 2,3 with values of 92,19 is too far from any existing spectral classes. But we have reached to maximum number of spectral classes specified by the user. Therefore pixel 2,3 will be assigned to the closest spectral class. The spectral distances are: to class #1: 86.8, to class #2: 181, to class #3: 113, and to class #4: 39.9. Therefore pixel 2,3 becomes a new member of class #4. Pixel 3,3 is also closest to class #4 with a spectral distance of 34.6. Pixel 3,3 is therefore forced to be a member of class #4. Pixel 4,3 is closest to class #2 with a spectral distance of 61. This pixels therefore becomes a member of class #2.

Spectral Class	Class Members	New Class Means
1	1,1	5,3
	2,1	6.0,3.5
	1,2	6.0,3.67
	2,2	**6.5,3.75**
2	3,1	210,150
	4,1	204,161.5
	4,3	**220.33,173.67**
3	3,2	55,123
	4,2	**57,126**
4	1,3	55,34
	2,3	73.5,26.5
	3,3	**62.0,27.33**

Pass #2: Assign final class membership for each pixel based on the closest spectral distance to final spectral class:

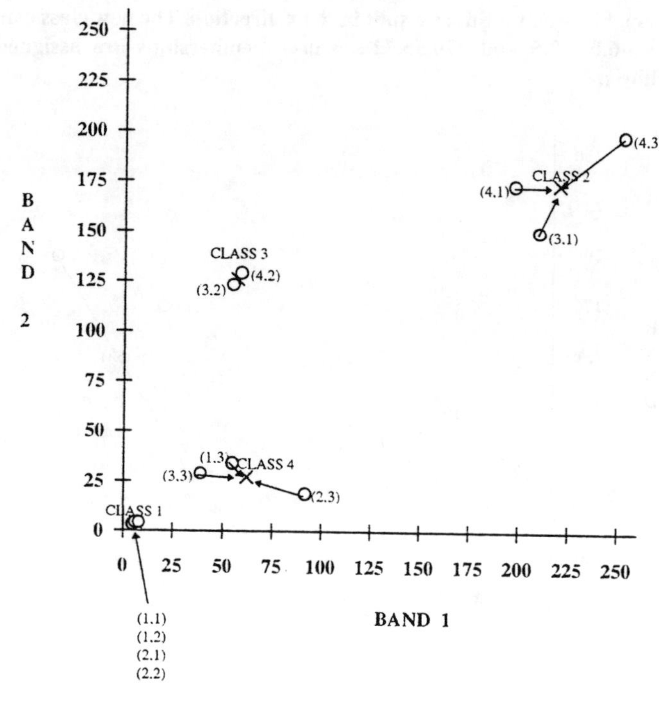

CLASSIFIED IMAGE:

<table>
<tr><td colspan="2" rowspan="2">BLACK</td><td>BROWN</td></tr>
<tr><td>YELLOW</td></tr>
<tr><td colspan="2">GREEN</td><td>BROWN</td></tr>
</table>

4) Perform an ISODATA unsupervised classification using bands 1 and 2 from question #1 using the following criteria:

Starting Criteria

Initial number of spectral classes: 1

Processing Criteria

Splitting criterion:

maximum variation of spectral class: 100

Merging criteria:

minimum distance between spectral class means: 25

maximum number of spectral classes: 3

minimum number of members in a class: 2

Stopping Criteria

Maximum number of iterations: 3

Desired percent unchanged class members: 90

Iteration #1: We have one spectral class of all 12 pixels. The means and standard deviations are: X = 82.25, SDx = 88.3 Y = 72.5, Sdy = 75.5. The standard deviations both exceed the maximum allowed (set as 10 by the user). Since Sdx > Sdy, we split in the x direction. The new class centers are –6.05,72.5 and 170.55,72.5. Class memberships are assigned as follows:

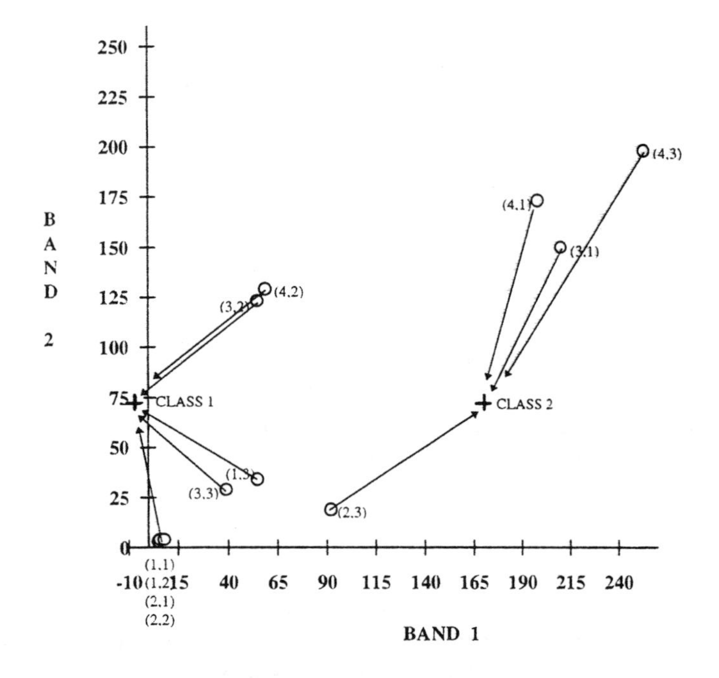

The new means and standard deviations are:

Spectral Class	Mean X	SDx	Mean Y	SDy
Class 1	29.3	25.0	41.3	53.7
Class 2	188.3	68.4	135.0	79.8

Check if any classes should be merged. Are there any spectral classes with a spectral distance between means less than 25? Are there more than three spectral classes? Are there spectral classes less than two members? No, so we do not merge any spectral classes.

Check stopping criteria. Is the number of iterations equal to three? Are the percent unchanged class members greater than 90%? No, so we continue with iteration #2.

Iteration #2: Check if any classes should be split. Both class 1 and class 2 should be split in the Y-direction because the standard deviations are greatest in that direction. New class centers are assigned as: X = 29.25, Y = –12.45 X = 29.25, Y = 94.95 X = 188.25 Y = 55.2 X = 188.25 Y = 214.8. Class membership is then assigned to each pixel based on closest spectral distance:

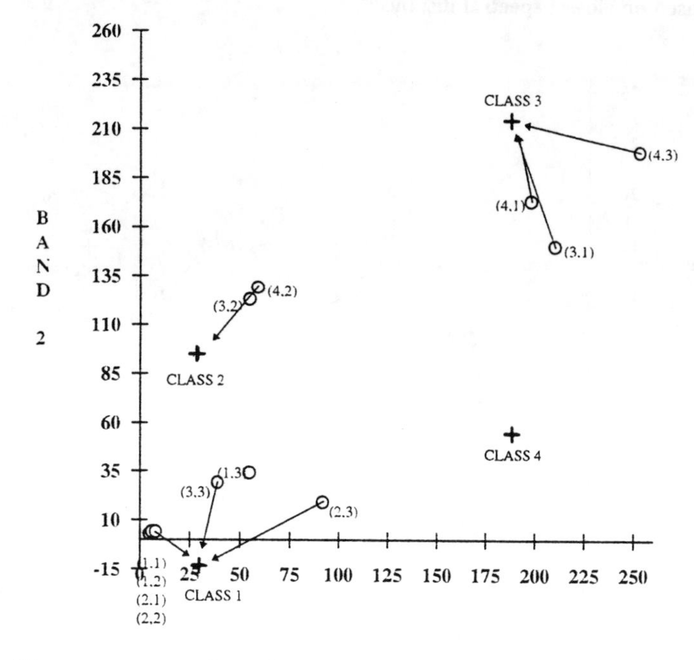

The new means and standard deviations are:

Spectral Class	Mean X	Sdx	Mean Y	SDy
Class 1	30.3	33.6	13.9	13.4
Class 2	57.0	2.8	126	4.2
Class 3	220.3	28.9	173.7	24.0

Check if any classes should be merged. Are there any spectral classes with a spectral distance between means less than 25? No. Are there more than three spectral classes? No. Are there spectral classes with less than 2 members? No.

Check stopping criteria. Is the number of iterations equal to three? Are the percent unchanged class members greater than 90%? No, so we continue with iteration #3.

Iteration #3. Check if any classes should be split. Both class 1 and class 3 should be split in the X-direction because the standard deviations are greatest in that direction. Class 2 has standard deviations less than the user-specified value of 10. New class centers are assigned as: X = –3.3, Y = 13.9 X = 63.9, Y = 13.9 X = 191.4 Y = 173.7 X = 249.2 Y = 173.7 and 57,126 (old class 2). Class membership is then assigned to each pixel based on closest spectral distance:

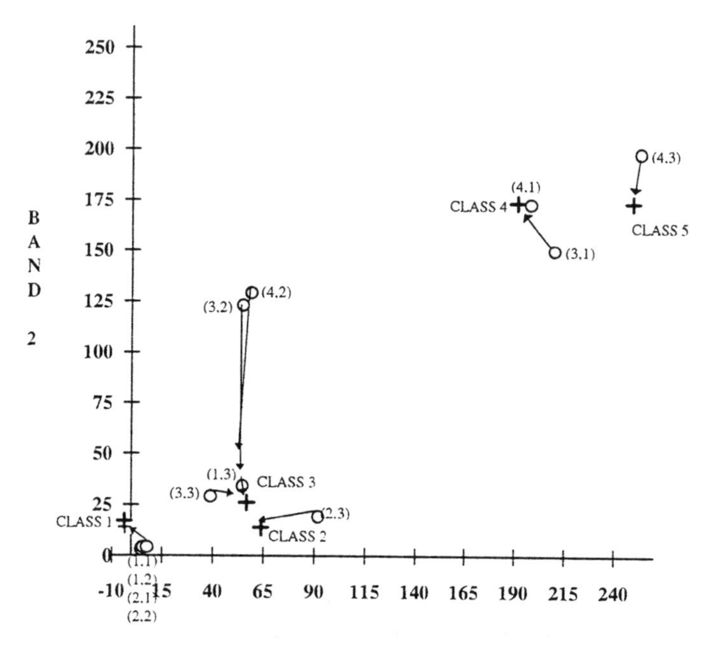

The new means and standard deviations are:

Spectral Class	Mean X	Sdx	Mean Y	SDy
Class 1	6.5	1.3	3.75	0.5
Class 2	92.0	0	19.0	0
Class 3	52.0	8.9	78.8	54.7
Class 4	204.0	8.5	161.5	16.3
Class 5	253.0	0	198.0	0

Check if any classes should be merged. Are there any spectral classes with a spectral distance between means less than 25? No. Are there more than three spectral classes? Yes, therefore merge the closest classes. The spectral distances between classes are:

Spectral Classes	Spectral Distance
1 to 2	86.8
2 to 3	71.9
4 to 5	61.1

Therefore classes 4,5 and 2,3 are merged. The three spectral classes are as follows:

Class	Members	Mean X	Sdx	Mean Y	Sdy
1	1,1 1,2 2,1 2,2	6.5	1.3	3.75	0.5
2	2,3 3,2 4,2 1,3 3,3	60.0	19.5	66.8	54.4
3	4,1 3,1 4,3	220.3	28.9	173.7	24.0

Are there spectral classes with less than 2 members? No.

Check stopping criteria. Is the number of iterations equal to three? Yes, so stop and proceed with pass #2.

Pass #2. Assign spectral class membership based on closest spectral distance:

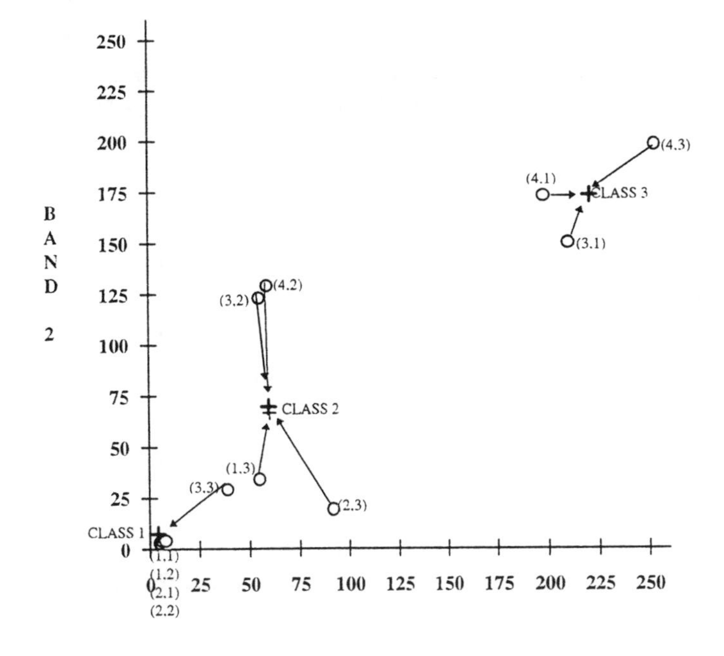

CLASSIFIED IMAGE:

1	1	3	3
1	1	2	2
2	2	1	3

CLASSIFIED IMAGE USING PALETTE FROM QUESTION #1

<table>
<tr><td colspan="2" rowspan="2">BLACK</td><td colspan="2">BROWN</td></tr>
<tr><td colspan="2">GREEN</td></tr>
<tr><td colspan="2">GREEN</td><td>BLACK</td><td>BROWN</td></tr>
</table>

CHAPTER 7

2) Use the min/max parallelepiped classifier to classify the same image as problem #1. Shade in the predicted burn pixels.

To be within a "Burned" parallelepiped, a pixel must have a band 1 value from 50 to 70, and a band 2 value from 50 to 70. To be within a "Unburned" parallelepiped, a pixel must have a band 1 value from 60 to 120, and a band 2 value from 65 to 120. Pixel 3,1 with values of 70,70 is within both parallelepipeds. Therefore use the nearest neighbor rule:

spectral distance to "Burned" means: 13.1 (assign pixel to this class)

spectral distance to "Unburned" means: 36.6

Pixel 2,4 with values of 60,60 is also within both parallelepipeds. Using the nearest neighbor rule, this pixel is assigned to the "Burned" class.

Original Image

80	100	70	50	55
70	70	70	60	55
80	70	100	80	60
90	90	90	70	60
55	60	65	110	110
55	50	60	90	100
55	60	60	100	120
60	60	80	110	110

Classified Image

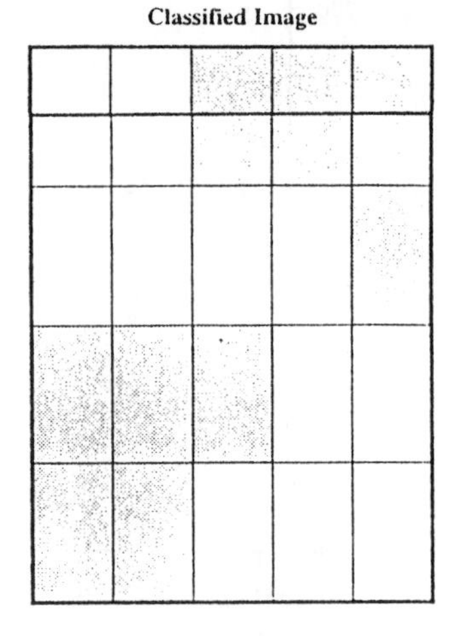

4) A waterfowl biologist wants a map of a marsh area where two vegetation types (cattail and smartweed) dominate. The mean digital values from representative training samples are:

Vegetation Type	Band#1 Mean	Band#2 Mean	Band#3 Mean	Band#4 Mean
Cattail	55	104	68	123
Smartweed	44	75	58	102

Classify the image using the minimum distance to means classifier. SHOW YOUR CALCULATIONS!

Pixel (Col, Row)	Cattail Spectral Distance	Smartweed Spectral Distance
1,1	SQRT($(55\text{-}50)^2 + (104\text{-}81)^2 + (68\text{-}59)^2 + (123\text{-}111)^2$) = 27.9	SQRT($(44\text{-}50)^2 + (75\text{-}81)^2 + (58\text{-}59)^2 + (102\text{-}111)^2$) = 12.4*
2,1	SQRT($(55\text{-}52)^2 + (104\text{-}77)^2 + (68\text{-}63)^2 + (123\text{-}120)^2$) = 27.8	SQRT($(44\text{-}52)^2 + (75\text{-}77)^2 + (58\text{-}63)^2 + (102\text{-}120)^2$) = 20.4*
1,2	SQRT($(55\text{-}49)^2 + (104\text{-}100)^2 + (68\text{-}59)^2 + (123\text{-}120)^2$) =11.9*	SQRT($(44\text{-}49)^2 + (75\text{-}100)^2 + (58\text{-}59)^2 + (102\text{-}120)^2$) = 31.2
2,2	SQRT($(55\text{-}50)^2 + (104\text{-}98)^2 + (68\text{-}61)^2 + (123\text{-}107)^2$) = 19.1*	SQRT($(44\text{-}50)^2 + (75\text{-}98)^2 + (58\text{-}61)^2 + (102\text{-}107)^2$) = 24.5
1,3	SQRT($(55\text{-}47)^2 + (104\text{-}79)^2 + (68\text{-}65)^2 + (123\text{-}108)^2$) = 30.4	SQRT($(44\text{-}47)^2 + (75\text{-}79)^2 + (58\text{-}65)^2 + (102\text{-}108)^2$) = 10.5*
2,3	SQRT($(55\text{-}49)^2 + (104\text{-}99)^2 + (68\text{-}63)^2 + (123\text{-}120)^2$) = 9.7*	SQRT($(44\text{-}49)^2 + (75\text{-}99)^2 + (58\text{-}63)^2 + (102\text{-}120)^2$) = 30.8

Color your classified image using the following color palette assignment:

Cattail → yellow

Smartweed → green

Original Image		Classified Image	
50	52	GREEN	GREEN
81	77		
59	63		
111	120		
49	50	YELLOW	YELLOW
100	98		
59	61		
120	107		
47	49	GREEN	YELLOW
79	99		
65	63		
108	120		

6) A watershed hydrologist is interested in classifying snow cover within a drainage basin. From training fields the following likelihood contours are developed:

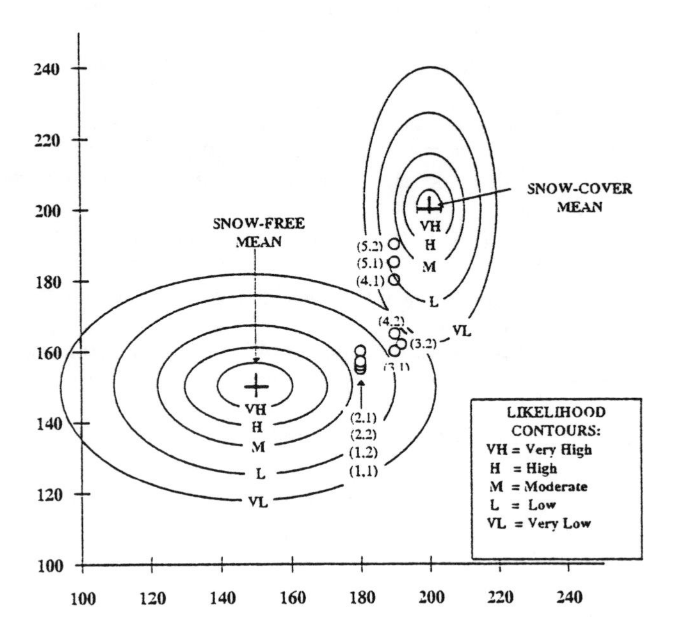

Using the maximum-likelihood rule, predict whether the following pixels are snow-covered or snow-free. Shade in the snow-free pixels.

Assign the likelihood contours the same values starting from the peak likelihood at the means location. Then plot the spectral location of each pixel.

From the plot above, we can see that the maximum likelihood for the snow-free class occurs with pixels 2,1 2,2 1,2 1,1 3,1 3,2 and 4,2. Shade in these pixels as predicted snow-free:

180	180	190	190	190
155	160	160	180	185
180	180	192	190	190
156	157	162	165	190

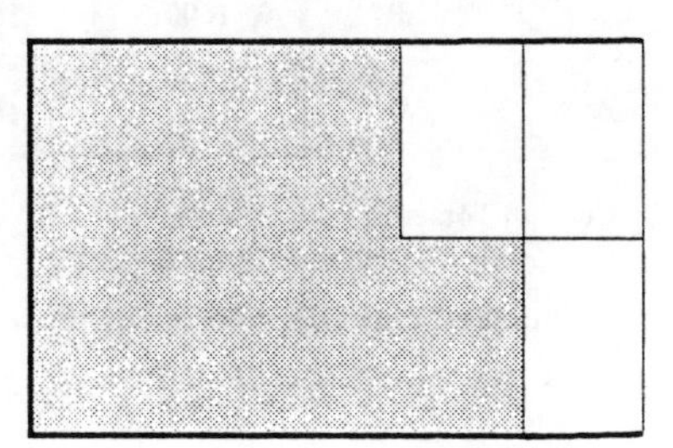

CHAPTER 8

2) Give an example of an error matrix that would have a KHAT value of 0.0.

Any matrix with each cell having the same value will have a KHAT value of zero.

"GROUND TRUTH"

CLASSIFIED	BH	SP	LP	W
BH	**10**	**10**	**10**	**10**
SP	**10**	**10**	**10**	**10**
LP	**10**	**10**	**10**	**10**
W	**10**	**10**	**10**	**10**

Overall classification accuracy = 40/160 = 0.25

Products of Row and Column Totals

	BH	SP	LP	W	Row Totals
BH	1600	1600	1600	1600	40
SP	1600	1600	1600	1600	40
LP	1600	1600	1600	1600	40
W	1600	1600	1600	1600	40
Column Totals	40	40	40	40	

Expected classification accuracy = (1600 + 1600 + 1600 + 1600) / 25,600

= 0.25

KHAT = (0.25 – 0.25) / (1 – 0.25) = 0.0

4) A wetlands scientist obtains a Landsat Thematic Mapper classification of a 10,000 hectare freshwater marsh. To assess the accuracy of the digital classification, an accurate vector GIS of ten randomly located 100-ha blocks is developed by interpreting color infrared aerial photographs taken the same day the Landsat TM image was acquired. The minimum mapping unit for photo interpretation was 2.5 acres (smallest polygon is 2.5 acres).

The overall classification accuracy of the digital classification was assessed as 85% using the GIS as reference data. Is this estimate likely to be conservative or optimistic? Explain your reasoning.

Conservative — the actual classification accuracy is likely to be greater than 85%. Because polygons with a minimum mapping unit area of 2.5 hectares were used, small scattered stands may have been correctly classified but would be evaluated as incorrect because the reference data was generalized to 2.5 hectare polygons.

6) A forester produces a digital classification of old growth forest. The accuracy of the classification is assessed independently by several groups. Match the following accuracy assessment scenarios with the most likely overall classification accuracy.

____ 50% overall classification accuracy
____ 75% overall classification accuracy
____ 90% overall classification accuracy

Group A: A guided approach was conducted in conjunction with the field work for training data collection, thereby significantly reducing the costs of data collection. A systematic sample in which the center of a polygon from the center of every 20th air photo plot was used as reference samples. To minimize non-classification errors (such as slight positional shifts) only homogeneous polygons of at least 5 hectares were used as reference data. The center pixel of these polygons was selected for each reference data sample location. The photos were taken the same month and year as the satellite image was acquired. **Likely to be an optimistic estimate — 90%.**

Group B: Using high altitude color infrared photographs (taken by NASA in 1979), we stereoscopically interpreted cover types to polygons with a minimum mapping unit of 2 hectares. The covertypes were transferred to the USGS 1:63,360 quadrangle series using a digitizing tablet and stored in a vector GIS. For accuracy assessment these GIS polygons were rasterized to correspond to the classified image. Classification accuracy assessment was conducted on a pixel by pixel basis between the reference data and the classified image for a test area of 1000 ha. **Likely to be a conservative estimate due to use of generalized polygons interpreted to some minimum mapping unit — 50%.**

Group C: All pixels that were classified as old growth were selected. A random sample of 100 classified old growth pixels then selected. For each selected pixel, we recorded the pixel UTM coordinates using a cell coordinate inquiry. We then programmed a 6-channel GPS to guide us to these UTM coordinates. The actual covertype was then recorded at these GPS-guided locations. **Probably slightly conservative due to inherent positional errors — 75%.**

Index